Preface

This book provides discussion on society's dependency on fossil fuels today, and how our culture is impacted by the continued reliance we have on this resource. The discussions and empirical examples provide insight to fossil fuel related tragedies and the impact to human health, environment and cultural identity; but also describe the beneficial products and economic impacts gained from the fossil fuel industry today. Initially, emphasis focuses on the detrimental impact from oil, coal and diesel fuels that have caused severe damage to human health, wildlife and the environment. Insight is provided on the socio-economic impacts experienced through current day and historic fossil fuel related tragedies. Details are provided which discuss the extraction, delivery and storage of fossil fuels, and the recurring associated tragic events. Hydraulic Fracturing "fracking" is discussed in the controversial relationship as a valuable extraction technology, but also as a cause of contamination and environmental damage.

In contrast to negative impacts, this book provides discussion on the beneficial aspects of fossil fuels and their importance to society in providing heat, cooling, and essential transportation fuels. The discussions utilize onsite reporting and scientific reference to show the significance of the substantial increase of natural gas and oil availability recently discovered in the U.S. shale basins. These new shale resources are shown to have a positive impact to the economy and increased availability of the domestic petroleum resources.

This book also discusses how national leadership, the petroleum industry and science have key responsibilities in seeking effective alternative energy solutions, and strengthening public awareness. Collaboration in alternative energy research and development is discussed as a key strategy in decreasing society's dependency on fossil fuels. Cooperative dialogue is shown to be a successful step and effective tool among all stakeholders.

Table of Contents

CHAPTER 1 3

   Fossil Fuel in U.S. Culture..3

CHAPTER 2 14

   Socio-Economic and Health Impacts from Fossil Fuel Disasters.................................14

CHAPTER 3 35

   The U.S. Natural Gas and Oil Boom: Abundant Resources in Shale Formations.......35

CHAPTER 4 51

   The Future of U.S. Energy: Leadership, Alternative Solutions and Collaborative Strategy 51

Final Remarks...62

References...65

CHAPTER 1

Fossil Fuel in U.S. Culture

The reliance that society has on fossil fuel, specifically oil and petroleum, has a significant impact to our daily lives, community cohesiveness, and in the traditions that we identify with cultural importance. Most of the petroleum based products, as well as the industry that supports fossil fuel delivery and consumption, are essential in our society. Fossil fuel products provide us with energy, transportation, economic benefits, and importantly employment opportunities. The Institute for Energy Research (IER) (2015) describes fossil fuels as products which "make modern life possible" (p. 1). It is important to recognize that the energy and products derived from fossil fuels are most beneficial to our way of life. Whether our fossil fuel products are derived from oil rigs in the ocean, transported through pipelines across land or imported from merchant vessels from South America or the Middle-East, it is clear that we depend on the products in everyday life. The same dependency is true for the coal and the natural gas deposits which are mined from far beneath the ground. The benefits derived from all of the fuel products provide society with a means of survival, and allow us to carry out everyday activities in comfort.

The Energy Information Administration (EIA) (2015) identifies the important petroleum-based products used today in the categories of "transportation fuels, oils for heating and electricity generation, asphalt and road oil and the feed-stocks to make chemicals and plastics" (p. 1). Further, EIA (2015) data states that about 76% of the 6.97 billion barrels of petroleum used in 2014 were gasoline, diesel and jet fuel. The IER (2015) recognizes that "of the fossil fuels, none has had more far-reaching effect on society than oil" (p. 1). The positive benefits that society gains from fossil fuels are immediately evident, but what is the trade off?

We enjoy homes, schools and workplaces that have heat, air conditioning and power with the flick of a switch. We have various modes of petroleum based transportation to choose from that easily transports a grandmother across town to the grocery store, an injured child to the hospital, or a business woman to the far reaches of the world. The reliance we have on fossil fuels definitely has impact on our lives in a myriad of ways, and it will for the foreseeable future. However, this dependency has also created detrimental impacts in many areas of society including pollution of our waters, air, coastlines and wilderness. Specifically, oil spills damage the ecosystem and environment and the many human cultures that are part of the locale where these disastrous events occur. Society's reliance on fossil fuels has reached a turning point where decisions for smarter energy sources have to be incorporated now, in addition to the current technologies we have in place such as solar, wind and geothermic systems. We can no longer settle for continued destruction caused from oil spills, the dependency of importing foreign petroleum, and the subsequent negative cultural impacts.

Fortunately, there are technologies available today which present tremendous alternative energy solutions which this book offers for consideration. Alternative systems such as geo-thermal energy which provide both cooling and heating from ground-source temperatures can be utilized to decrease our dependency on fossil fuel. Our society must continue to move toward harnessing these alternative energy sources and continue research and planning toward more-advanced technologies to decrease our reliance on fossil fuels. When society makes significant progress away from the dependency on fossil fuels, including imported petroleum, then our cultures will be much better preserved, protected, and we can sustain the importance and integrity of our history, traditions, health and environment.

Further data reported by the EIA (2012) states that the U.S. is dependent on 18.8 million barrels of oil each day, which makes us the world's largest consumer of petroleum, and in 2011 about 45% of the crude and petroleum products consumed in the U.S. came from foreign imports. However, our local history and way of life can be altered due to the impact of our reliance on fossil fuel, whether it is resulting from the processes of extraction and delivery, random oil spills (large and small) or in the resulting waste and waste by-products. Additionally, subsequent health related problems result from our reliance on fossil fuels in both humans and creatures.

David Biello (2010) discusses specific harmful (cultural) impacts related to fossil fuels, specifically resulting from oil spills, and provides insightful comparative analyses of historic spills and the damage to sea, fish, shores and wildlife. Biello (2010) points out how polycyclic aromatic hydrocarbons (PAHs) which are among the most dangerous toxins in oil, causes sickness in humans, wildlife and plants when ingested or inhaled. Significantly (and a horrifying thought to clean-up crews) these PAHs can have long-lasting damage once in the body, including affecting our DNA. Evaporation does help to decrease 20 to 40 percent of the PAHs once the oil reaches the surface, however plumes of oil which remain under the water can extend for several miles and toxic compounds are washed off and contaminate the water. This contamination finds ways into the salt marshes and harms the wildlife. The *Exxon Valdez* oil disaster in 1989 still affects the local sea otters as they dig up oil while searching for clams in Prince William Sound. Similar damage may impact the 16,000 species of plants and animals that live in the Gulf of Mexico where millions of gallons of crude spilled in 2010 from BP's damaged oil well (Biello, 2010).

This impact to human health is incurred during clean-up efforts where oil covers the skin and clothing of the workers, and also by the fumes that are inhaled during cleanup efforts. Additionally, the fumes from spilled petroleum saturate neighborhoods adjacent to spill areas causing further sickness and stress among humans. Antonio Juhasz (2012) discusses sickness which has affected people in the Gulf of Mexico two years after the oil spill. Juhasz (2012) conducted an interview with Nicole and William Maurer, residents of Buras, Louisiana, during which Nicole described how the family has been plagued by health problems. After the disaster, the smell of oil was persistent and heavy for months around the home of the Maurer family. The health problems described are consistent with illnesses associated with petroleum spills including upper respiratory infections, breathing difficulties, rashes, upset stomachs, etc., all of which according to Nicole, never seem to end. The exposure was significant for William who was part of the Vessels of Opportunity clean-up program, and would come home daily with his clothes and body smelling of oil. Within six months William had started bleeding from his ears and nose, and developed a persistent heavy cough.

William's work had also suffered greatly due to the loss of fishing which sustained the family's income. Nicole relates that she believes that the legal settlement is not going to bring justice for the family's health and exposure problems, and makes a sobering comment in that she wants just one thing; enough money to get her family out of the Gulf Coast for good (Juhasz, 2012). This instance of distressed impact to the health and income of family is a critical sign in that it validates the negative impact to culture as these disastrous incidents cause people to migrate from their livelihood and homes. This book will provide further insight to the various fossil fuel disasters experienced in the U.S. and present discussion on resulting damages to the health of humans, wildlife, and how this impacts the effected society's way of life.

While this book was being developed, yet another oil spill caused disruption in the Mississippi River. The U.S. Coast Guard's Unified Command (2013) reported a crude oil spill that occurred in the lower Mississippi River, near Vicksburg, MS, where two tank barges collided into a bridge, resulting in damage. One barge contained more than 80,000 gallons of crude, which was releasing oil into the river, as containment crews attempted to defer the damage. In this incident, as with all others on the river, the Coast Guard closed the maritime traffic resulting in vessels lining up north and southbound for at least 19 miles (U.S. Coast Guard Unified Command, 2013).

In separate news coverage of this incident, Karl Plume (2013) brought attention to the importance to the shipping industry's daily use of the Mississippi River and the associated inland waterways for transport of goods. These vessels depend on the river and waterways to haul billions of dollars in grain, coal, fertilizer and other commodities every year (Plume, 2013). This crude oil spill from the barges not only disrupted the continued navigation and commerce along this waterway, but also threatened the environment, drinking water and the safety of the neighboring communities.

Dr's Kirstin Dow, Seth Tuler and Thomas Webler (2010) of the Social and Environmental Research Institute (SERI) have developed the detailed research project entitled "Human Dimensions Impacts of Oil Spills" which provides exceptional insight and empirical data on the effects of oil spills and the impact to social, cultural and health in a community. A section of this project, *Impacts Associated with Municipal Services,* discusses the collision of the oil tanker *T/V Tintomara* and fuel barge *DM-932* near downtown New Orleans, Louisiana. The article highlights how oil spills severely impact the maritime transport, businesses and industries that are part of the overarching distribution and transportation sector (Dow et al., 2010). In this case, the barge ruptured in half and 282,828 gallons of oil spilled into the Mississippi causing pollution along approximately 100 miles of the River, leaving towns located downriver impacted by loss of drinking water for more than 18 hours. Upon reopening of the water valves, the water still had to go through a cleansing process further prolonging the ability to drink and use potable water. Still, following cleansing, residents in the areas were skeptical about drinking the water and in this incident alone the economic impact for all ports affected by the oil spill was estimated at $275 million a day (Dow et al., 2010).

These disruptions to daily life are severe and many of the damaging and lasting impacts of fossil fuel incidents are not readily appreciated by those outside of the community. As we see so often, an event with such devastating impact is soon forgotten as the media moves onto the next "big thing", eventually taking the attention away from the loss felt by those "distant" communities; loss that will be experienced for decades, or perhaps forever depending on the scale of the event. While the news of the destructive events subsides, the devastation is no longer given much attention by the public at large. This is apparent with the 2010 oil spill in the Gulf of Mexico. Once the pipe was capped the media switched to covering other aspects of the story such as the developing litigation against the BP Corporation, and then eventually moved onto other news stories not associated with the Gulf disaster. In the aftermath of this particular disaster, the focus moved to advertising campaigns emphasizing a healing Gulf where tourists would be welcomed with clean beaches and fresh untainted seafood - fun for everyone. Thus the negative implications of fossil fuel disasters are soon forgotten in the outer public eye - until the next spill. However, the local community leaders have no choice but to advertise that everything is fine because they are fighting for the livelihood of their communities; communities that still harbor the negative impacts of loss of life, jobs, pollution and lingering health problems from the spill. It is essential for these leaders to bring back tourism quickly in the attempt to regain economic strength in the most affected areas.

The reliance our society has on fossil fuels is powerful, and even with the many devastating impacts such as the oil spills, we continue to welcome the products that oil and petroleum have to offer. Our reliance on this resource is inherent and we have evolved into this heavy dependence over centuries – as oil has historic relevance in societies throughout the world. The IER (2013) discussed the use of oil in architecture, ship caulks, medicine and roads in Mesopotamia as far back as 3000 BC. Additionally, the IER discussion shows that in ancient China 2000 years later, crude oil was used in lamps and for heating of homes. Additionally, in the late 19th century oil made a significant impact in the U.S. by displacing whale oil as fuel for lighting. Petroleum became the prominent fuel for the automobile with the development of the internal combustion engine, as it still stands today (Institute for Energy Research, 2013). Fossil fuel has indeed provided significant impacts to the comfort of mankind. But what ultimate price do we pay? What hazards does our society incur by the continued dependency on fossil fuels?

Dow et al. (2010) provide a very important look at how the culture, or traditions and social cohesion, of a community can be altered as evidenced in Buzzards Bay Massachusetts. The authors describe this area is an extensive development of eleven communities along 350 miles of coastlines, with harbors, beaches, parks and private land. In this incident of 2003, a tank barge, the *Bouchard-120* went aground and spilled 98,000 gallons of oil into the Bay. The oil spill severely affected the aesthetic value of the area, with black tar balls, and lasting images of pollution. In this case, the people of this rural community of Buzzards Bay are tightly knit, and being located away from larger regions gives them a sense of security from outside problems. The spill of the tank barge not only caused damage to large areas of the water and land, but significantly to the rocks in the popular public beach areas where families and friends have gathered and enjoyed for generations. Importantly, the oil spill covered a particular swimming area and rock that has been an icon for generations of families, in which children's ability to learn to swim to the rock was a recognized community milestone in growing up (Dow et al., 2010). The residents claimed that "fouling of the rock tainted a rite of passage for local children and was a blow to the identity of the town", adding that the cultural impacts to small communities can be very pronounced, affecting the way of life for the people of which traditions of the region define their identity (Dow et al., 2010) (p. 5). This incident, and many like it, has damaged pieces of the community identity and important family memories. In the cases of oil spills, whether these are along a coast line, river or bay, it is not so easy for the community to rally and fix the damage which has taken place, and in many cases traditions and aesthetic community scenes that generations have enjoyed with their families for decades will never be enjoyed again.

An additional impact to cultures is the forced displacement of people, or entire communities due to fossil fuel related accidents. Population displacement and forced migration is an effect caused by environmental changes which impact a culture or community in a significant way. Specifically when environment and population is affected by oil spills or other incidents, communities and larger populations have to abandon their settlements where they make their livelihood, either temporarily or indefinitely. McLeman (2011) discusses studies of settlement abandonment, and the interactions between the environmental and non-environmental factors of abandonment, and provides insight on the entailing significant costs and disruption that occur with such migration. A single environmental event can devastate a small community or region, causing temporary or permanent displacement of individuals. This is particularly evident when the area is contaminated to a degree where air, water or subsistence resources become toxic.

An example of a single environmental event from fossil fuel includes the coal mining community of Centralia, Pennsylvania that was devastated by an underground coal fire. This underground inferno fueled by coal crept through the area over many years (1989 till present day) and resulted in permanent abandonment. McLeman (2011) offers another example in the Prypiat (Chernobyl) nuclear accident in the former Soviet Union. This was a man-made disaster in which the sudden onset event necessitated permanent evacuation of the affected area (though a few individuals have remained). McLeman (2011) provides the insight that migration can be temporary or indefinite, caused by natural events or manmade, influenced by a multitude of social, economic, and environmental factors.

The realization that fossil fuel related accidents can initiate any degree of migration, or displacement of people from their long-lived community, with established cultural identities, traditions and life-style is most significant. This awareness provides a lens for society to think about humanity's ability to maintain habitation in their original geographical location, and the range of events set in motion culturally due to forced migration. Forced migration, whether it is temporary or permanent, has a detrimental impact to the societies affected. This is especially relevant because these disastrous events continue to take place and have for decades. This is particularly represented in great extent in the April 2010 explosion of the *Deepwater Horizon* rig disaster where over five million barrels of oil was released into the sea before it was able to be capped. Also, this is represented in the 1989 *Exxon Valdez* oil spill where an estimated 11 million gallons of crude oil was released into in the Prince William Sound. In both of these cases there was extensive damage to sea, environment, and human culture. The families and businesses along the Gulf are still assessing the permanent damages, while the Native Americans along the Prince William Sound have realized for over two decades the permanent impacts to their culture, way of life and the environment.

Today we are witnessing the evolution of negative impacts to the Gulf residents from the *Deepwater Horizon* oil spill. Barcott (2010) provides on-sight reporting of the environmental and socioeconomic impacts, particularly in the Louisiana wetlands. This area where the Mississippi River Delta meets the sea is a vast geographic of, "barrier islands, beachheads, open bays, canals, marshes and freshwater swamps running inland for 25 to a hundred miles" (Barcott, 2010, para. 1). Barcott provides visual descriptions of oil's negative effect during its movement into the bayou consisting of about 12,355 square miles of ecosystem. While the oil's destruction was significant to the ecosystem, fish, birds and other wildlife, it also had a heavy impact on family fishing industries. In this case the devastation was to the socioeconomics of the Barataria-Terrebonne estuary where families relied on the annual fishing and crabbing. The impact also produced associated impacts on businesses such as the closing of restaurants and fishing supply stores (Barcott, 2010). This book will discuss in more detail, in following chapters, how the socio-economic, health and welfare of humans, and the environment has been impacted by fossil fuel disasters.

Once we have an understanding of the impacts of the energy sources in our lives, particularly the fossil fuels, we can put an analytical lens to the problem of dependency. At the same time, future directions we might take to lessen the negative impacts to our culture can be analyzed. It is important that we see the overall commitment our society has in the use of fossil fuels including the requirements we have for importing that 45% from foreign sources. We should understand and give proper recognition to the cultural loss suffered in every incident.

Loss of traditions, cultural identification, or skills learned over time all passed down from generation to generation is at stake from the negative impacts of reliance on fossil fuels. While our society does enjoy the benefits from energy and petroleum products, the price paid from oil spills is significantly high, and not an acceptable trade off, including the losses suffered in a limited locale which are just as severe to a smaller cultural footprint. The example of the oil spill which caused irreversible damage to the small community of Buzzards Bay, MA (Dow et al., 2010), is a scenario that anyone in the U.S. can apply to their own locale.

Traditions and community bonding are an important part of life; they provide a history and anticipation of the future as families and communities continue to grow. When these identities are taken away due to oil spills, or fossil fuel wastes, it is not only a detriment to the teachings of a new generation, but also to past generations. Loss of tradition and artifacts is a sad and unfortunate circumstance, particularly when such impairment is from disastrous events. Our customs provide the link to our beginnings, and protect the lessons handed down from family members who have since passed on from this life. When our ways of life are damaged due to the contamination from fossil fuel, and the aesthetics of our homestead and relics are lost, then our historic identity is altered. When subsistence is destroyed, such as loss of fishing areas and wildlife, we lose important pieces of our livelihood, and skills passed down for decades; and it impacts greatly the social unity in a community. This book identifies the importance that fossil fuels have in our society, providing many beneficial products, comforts, jobs and business opportunities. However, as the discussion indicates there is much negative impact from this reliance on fossil fuels that we must be keenly aware. We must incorporate the insight of the harmful consequences from use of fossil fuels into future planning, specifically in the protection of culture, health and in environmental preservation.

CHAPTER 2

Socio-Economic and Health Impacts from Fossil Fuel Disasters

Damage and contamination from fossil fuel spills and the related fossil wastes have lasting and harmful effects to regional economics, human health, wildlife and our environment. For humans, the fossil fuel accidents can cause temporary or permanent job losses, physical and stress related health problems, and property damage. The harmful effects to sea life and wildlife are experienced immediately, as spills and wastes damage the health of all species and their habitats, reducing their ability to spawn, reproduce and find refuge, and ultimately degrading the food web.

Research identifies a multitude of cases where fossil fuels have caused severe damage to communities and to the health and welfare of the human and wildlife populations. The following discussion provides current and historical empirical examples that have occurred as a result of fossil fuel related incidents. The details in these cases provide further evidence that society's continued dependency on fossil fuels has damaging impacts on our way of life and health. As well, the impacts are severe to the wildlife, ocean life, the ecosystems and natural habitats. Additionally, the uncontrolled factors described in these situations validate the critical reality that alternative energy technologies and sources must continue to be improved. And, these technologies must continue to be researched and applied, in order to move our society ahead, and away from the current levels of fossil fuel dependency.

To initiate a proper perspective, the two most influential fossil fuel related disasters in the history of the U.S. should be forefront in this chapter's discussion. These incidents would be the crude oil spills associated with the *Deepwater Horizon* in 2010 and the historical *Exxon Valdez* in 1989. These two disasters separated by over two decades, and by several thousand miles, provide an ideal lens in which to analyze the severe impact to the societies involved. The severity of the *Deepwater Horizon* oil spill covers an extensive region of coastline and fishing grounds in the Gulf of Mexico. And, tragically this incident included the loss of eleven men, and injury of seventeen others, as a result of the initial explosions on the drilling rig. Cultures affected from this disaster consist of generations of French settlers (Cajuns), Vietnamese and Croatian immigrants. Overall impacts of this incident are still being assessed, and recovery, in all realms, is an ongoing process. The *Exxon Valdez* oil spill provides a historical model and exceptional critical data of the negative impacts incurred by societies of a large region, specifically with regard to health, socioeconomics and wildlife. The large Native Alaskan population was affected, and their dependence on the fishing and wildlife resources was a specific injury incurred. Twenty-four years later, the impacts from the *Exxon Valdez* incident are still being assessed and the recovery is still ongoing in many areas. The following provides a closer look at the social impacts from these two disasters, separated by time, geographic location and cultural differences.

The *Deepwater Horizon* oil spill represents the biggest man-made disaster in U.S. history, impacting the coastal littoral of the northern Gulf of Mexico. The oil spill shut down fishing areas from off the coast of Texas, all the way over to the panhandle of Florida. The crabbing and oyster grounds in the marshes and inlets of Louisiana were shut down, effectively halting the daily life along the coastal communities. Income and trade was at a standstill with no idea of how long, or to what level of degree this disaster would take – this was a socioeconomic catastrophe. The small businesses that cater to the fishing industry, the grocery stores which supported small communities, and the large businesses such as hotels and restaurants all suffered some degree of economic damage from this incident. Additionally, the families that relied on employment from the oil industry and the 3,858 oil and gas platforms in the Gulf were left without work, with no indication of returning to work while the offshore drilling was halted.

The economic loss is estimated to be in the billions of dollars. Fisk (2013), of Bloomberg News, reports on the amount of settlements between the plaintiffs, the U.S. Government, and the British Petroleum (BP) company. BP agreed to settle on 8.5 billion dollars in lawsuits in 2012, in addition to 4.5 billion dollars in criminal penalties. Also, as the third anniversary of the oil spill passed in April 2013, BP continued to face new lawsuits by claimants who had filed just prior to the three year statute of limitations deadline. The dollar amount for overall damages appears to be indefinite for now, and BP could still face up to 17 billion dollars in fines from violations in the U.S. Clean Water Act (Fisk, 2013).

While the industries and businesses along the Gulf's coastal littoral were at an indefinite standstill, individuals sought other avenues of income, to little avail. This was especially true for those in the seafood industry. The disruption of the seafood industry had a cascading effect on a web of Gulf businesses. Many of the unemployed joined the clean-up efforts under the *Vessel of Opportunities* program, which helped to provide some level of income. Yet, some were not able to gain employment from the clean-up opportunities, and were left to worry about their livelihood, and how they would pay their bills and provide support to their families. The loss of work, and the inability to gain support from the work-related cleanup efforts was devastating as well, just as the people had also experienced after the Exxon Valdez oil spill. The lack of work, and limited opportunities in the clean-up jobs, creates despairing outlooks for many, and disrupts the entire society as personal income shrinks and bills pile up.

When the ability to sustain livelihood is halted, there is an instantaneous feeling of loss, confusion and fear. This worry and desperation contributes to mental stress and health related problems of the people affected by the disaster. The Center for Disease Control and Prevention (2010), reports on the behavioral health risks from the *Deepwater Horizon* spill, and states that stress from disaster is a primary element to increased health problems, both mental and physical. Significant stress can develop into depression, anxiety and substance abuse. The Substance Abuse and Mental Health Services Administration (SAMHSA) (2013), reports on behavioral health problems from the *Deepwater Horizon* oil spill, specifically for residents of counties in Alabama, Florida, Louisiana, and Mississippi. SAMHSA provided data collected from a National Survey on Drug Use and Health (NSDUH) of U.S. civilian populations aged 12 or older. The data "compared measures of substance use and mental health in the past month and past year for a time period prior to the oil spill (2007 to 2009), to a time period after the oil spill (2011)" (p. 1). The NSDUH found that populations in the Gulf Coast Disaster Area (GCDA) had increased illicit drug use compared to pre-spill periods, greater than the increase of illicit drug, and marijuana use in the remainder of the U.S. The survey reported that marijuana use had increased in the 26 year or older population. Prevalence of alcohol use was also increased for ages 12 and older, which was similar to the same periods in the remainder of the U.S.

However in the NSDUH survey, no increase was observed in the rates of psychological distress, mental illness, suicide attempts or the use of mental health services in the GCDA population. The SAMHSA observes that the non-substantial changes noted in the behavioral health after the spill, could be because some of the population sampled was in regions with less intense impact from the oil spill. Additionally, the survey targeted age groups rather than demographic groups. The SAMHSA states that further analysis on demographic groups, such as those industries impacted by the oil spill, could provide a more targeted focus on behavioral health changes (NSDUH, 2013). In this survey, the wide region of sampling appears to draw attention away from the groups who experienced more-direct impact of the disaster. A more targeted survey of the fisherman, families and small business owners in the sub-population, would present more realistic details on changes in mental and health problems. The results and assessment by the SAMHSA indicate that negative mental health impacts are still undetermined for specific groups.

Though mental health demographics are still being assessed, the economic and cultural impacts were devastating, particularly to the smaller communities inland. A closer look at the Barataria-Terrebonne estuary located in the Louisiana wetlands provides an excellent example of how massive oil spills and contamination can devastate local economy. Barcott (2010) describes the importance of the wetlands to the ecosystem, wildlife and the human habitat along this Louisiana coast, where one-third of the U.S. oyster and shrimp crop originates. This coastal littoral harvests 98 percent of the fish, shrimp, crab and oysters which depends on the habitat around the Barataria-Terrebonne estuary. The marshes, covering approximately four million acres south and west of New Orleans and bordering the Atchafalaya and Mississippi Rivers are a critical element to the seafood harvests of the entire region. Barcott (2010) quotes a local author, naturalist and resident of this region who stated that "these are working wetlands...the land, the wildlife, and the people are inseparable here" (para. 4).

Barcott (2010) provides onsite witness accounts of the damages to the fishing industry and the communities as the oil continued inland. Two months following the spill, the oil made its way into the marshes, striking the barrier islands and continued to make its way along the currents and sticking to the aquatic and shoreline vegetation. As the oil continued to move inland, oyster harvesters and shrimpers tried to harvest what they could before the fishing grounds were closed by the health department. The fishermen and oystermen arose early and worked hard during pre-dawn hours to salvage as much of the catch as possible.

Eventually, the regional fishing areas were closed, and the cities along the coastline suffered the associated economic losses. Restaurants and stores once full of customers were empty, and signs of protest against British Petroleum (BP) began cropping up on lawns and in the front of businesses. Barcott (2010) describes one of the large empty shrimp lots where vendors had gathered to sell buckets of fresh-caught seafood to locals. There was no shrimp to sell, and one vendor remarked that even if there were shrimp available no-one would be buying because of the fear of the toxic oil and the dispersants being used in the clean-up.

As observed in many areas affected by disaster, confusion and fear sets in within the population, with individuals lashing out against those believed to be at fault, and at the government entities who may not be providing evidence of effective leadership. Barcott (2010) observed the destruction caused by the overall incident, and stated that "without the marshes, the rich human culture of the bayou has no foundation" (Barcott, 2010, para. 3). Barcott was absolutely exact in his observation of the cultural damages in the Louisiana wetlands. The oil spill's impact was relentless to the people who live off the land and the freshwater resources there as it moved from the sea, into the back marshes of the inland communities that have survived for generations. The oil systematically attacked the livelihood of fisherman, families and the small businesses supporting their parishes. Daily routine and work was threatened as the situation quickly overtook their lives. Worry and panic increased with every inch that the oil moved. As the fisherman quickly worked to gather their catches, their ability to pay bills, maintain their equipment, buy groceries and support their families was diminished. The people who live in the wetlands depend on a healthy ecosystem for subsistence and livelihood, and as the oil continued its destruction, the cultural identity of the people was being corrupted.

Many ethnic cultures including the Cajuns, Vietnamese, Croatian and immigrants from the Canary Islands have built their lives in the wetlands of Louisiana. These people, all with their own stories of survival, had assimilated into these coastal territories, making this region their home. For generations, these people have developed their traditions and ways of life in the coastal wetlands. As well, the Native American tribes have lived in these marshlands for centuries, living off the land and harvesting the seafood.

The societies in this area have survived war, erosion from industrial modernization and hurricanes, and of course the damaging encroachment of the gas and oil industry. The fossil fuel spill has damaged these people's homes and livelihood. Courselle (2010) describes the cultures of the communities, each contributing ethnic traditions, culinary seafood dishes and original music genre to the region. Courselle comments that the damages from the toxic oil and dispersants that are introduced into these wetlands put these people's ethnic identities and livelihood at risk; risk of the loss of culture that cannot be repaired or regenerated from payments made by the oil companies.

Over 20 years before *Deepwater*, the U.S. witnessed another type of devastating oil spill. In March of 1989, the 987 foot merchant tanker *Exxon Valdez,* transporting North Slope crude oil to Long Beach, California, grounded at Bligh Reef, spilling approximately 11 million gallons of crude into the Prince William Sound on the southern coastal region of Alaska. The oil immediately contaminated the site of the spill, and quickly made its way down the southern Alaskan coastline. Within seven days, the oil had travelled 90 miles, and within 56 days the oil had traveled 470 miles along the Alaskan Peninsula. The detrimental economic impact to commercial fishing was realized immediately, with closures of the fishing grounds due to the oil contamination. The economy is still affected today, over two decades after the incident. The herring fishery suffered severely and was closed in the Prince William Sound for 15 years following the spill. And, while many of the other fish and shellfish are considered to be recovering, the pacific herring are in the category of not recovering.

The Exxon Valdez Oil Spill Trustee Council Update (2010) reported that 20 years later, oil still lingered in areas of the spill, and that oil in the subsurface still impacted the environment, wilderness and human recreation. Though tourism is recovering, oil is still seen in spots on beaches and in other areas, and visitors have reported seeing fewer of the wildlife, particularly the many species of birds and mammals. Five site surveys were conducted in 1994 and again in 1999 along the Kenai and Katmai coastal areas. These surveys provided a documentation of the lingering oil in various surface and subsurface areas. The latest recorded follow-on visits to these sites in 2005 found that oil on the surface areas had declined, however the subsurface oil persisted in amounts and consistency similar to the samples surveyed in 1999. Significantly the 2005 survey found that subsurface oil was "similar to samples collected 11 days after the spill" (Exxon Valdez Oil Spill Trustee Council Update, 2010, p. 18).

The oil spill's impact was detrimental to those who relied on the fish and wildlife for subsistence, especially the population of 2,200 Native Alaskans. The Exxon Valdez Oil Spill Trustee Council Update (2010) stated that "Subsistence is a central way of life for those affected by the spill, and the value of subsistence cannot be measured by harvest alone…the lifestyle encompasses a cultural value of traditional and customary use of natural resources" (p. 42). Additionally, the Trustee Council notes a 2004 survey of the communities involved, which states 83 percent of those responding believed that their traditional way of life had been damaged, and many of those believed that there had been no recovery. The Native population relied considerably on the harvests of fish and shellfish, seals, deer and waterfowl, and the "oil from the spill disrupted subsistence activities for the villages and approximately 13,000 other subsistence permit holders in the area" (Exxon Valdez Oil Spill Trustee Council Update, 2010, p. 41).

The damage to subsistence availability is one of the immediate results from fossil fuel related incidents, stemming from injured or dead fish and wildlife, the closure of fishing and hunting grounds, fear of contamination and toxicity of the food. As well, this damage subsequently causes the disruption of community lifestyle and customs, some of which are permanently altered, as observed with the Alaskan Natives. Dr's Dow, Tuler and Webler (2010) discuss the cultural impacts relating to subsistence loss for the Alaskan Natives, noting how this population relies on the broad resources available in these damaged habitats. The impacts are detrimental overall, but specifically when livelihood is threatened, then a society's way of life is altered considerably. Dow et al. (2010) describe that "subsistence harvesting maintains kinship and social cohesion, and is one of the markers that help Native people to define themselves" (p. 1).

Social cohesion includes community participation in subsistence activities which provide teaching of skills, telling of stories about life, which strengthens the bonds of family and society (Dow et al., 2010). The dependence upon the crude from the Trans Alaska Pipeline, mixed with the impending potential for human or mechanical error provided the perfect opportunity for disaster; a disaster which would destroy rich elements of this region's human culture and natural environment. The Alaskan Natives suffered permanent damage far beyond that of the temporary loss of fishing and hunting grounds; they suffered loss of their day to day routines, social togetherness and values.

Gill, Duane and Picou (2012) describe the Native Alaskan fishing communities, and the people there who have "strong economic, social and cultural ties to renewable resources – particularly, fishery resources damaged by the oil spill" (p. 4). And, that the oil spill significantly disrupted the diversity of the fishing related work and the "ethnic and subsistence lifestyles" (Gill et al., p. 4). Dow et al. (2010) describe the long lasting cultural impacts for the Native Alaskans where the severe damage to the marine resources caused disruption to the "cultural calendar of resource cycle availability, and tradition" (p. 1). The daily traditions such as gathering and repairing nets and fishing equipment, rigging boats and harvesting fish from the Sound were abruptly halted by the oil spill. This incident interrupted and altered life routines such as family gathering after a long day's work, and the telling of stories at the nightly supper table. And, planning for the next day's work, or the anticipation of the season's harvest were no longer the focus, as the oil and the outsider activity overtook their traditional conversations, and the sharing between friends and family.

The oil spill and cleanup efforts along the southern Alaskan coast also caused physical and mental health problems to the communities affected. Picou (2009) discussed that the fishing communities and residents of Prince William Sound were suddenly impacted by "massive social shock" (p. 6), as their home was flooded by federal and state officials, cleanup crews, and reporters. As the massive cleanup efforts were initiated, crime increased and the mental health departments were overwhelmed. Additionally, the social resources such as garbage and sewer collection were disrupted, and not able to properly support the community needs. Picou (2009) references a study of commercial fisherman and Alaska Natives in the Cordova community, which indicated that psychological distress problems persisted through 1995. The study surmised that "four years after the oil spill, sociological research clearly documented that the fishermen and Natives were at high risk for chronic social and psychological impacts, and were intimately linked to the contaminated waters", and that "the Alaska Natives were linked culturally through their various seasonal subsistence harvests" (Picou, p. 6). Picou (2009) noted that further support was provided to the community with education and social relations programs put into place during 1996 and 1997, however the mental health issues persisted.

The lessons and observations from *Exxon Valdez* and *Deepwater Horizon* can be applied not only to other coastal regions, but also to any part of the U.S. where spills and contamination from fossil wastes have occurred, or have the potential to occur. The U.S. has access to innumerable fossil fuel sources supported by an immense web of pipelines, ships and barges, and wells. These tragedies are prominent in our past, current day, and will be for the foreseeable future.

The coal production, storage and waste disposal in the U.S. is yet another fossil fuel industry that has demonstrated negative impacts to society. Besides the obvious concerns of air pollution from burning the coal, the waste by-products also affect humans and environment. On 22 December 2008 the coal powered plant in Kingston, Tennessee had a failure of one of the containment walls, releasing approximately 5.4 million cubic yards (one billion gallons) of coal ash from a containment area. In *Civil Engineering ne*ws, Landers (2009) states that the Kingston fossil plant provides about 10 billion Kilowatts of electricity annually and burns approximately 14,000 tons of coal each day. Landers described the devastation of this impact to approximately 275 acres of neighboring land and water. This incident resulted in a massive slide of coal ash sludge, four to six feet deep, across the almost 300 acres, covering roads and filling the coves and boat docks that led up the properties of people living on the River. No physical human injuries occurred, however at least 40 homes were affected, with at least three having to be condemned. The ash spill entered the Emory River, a tributary of the Tennessee River, and threatened the water supply in areas where the sludge was most prevalent.

The Tennessee Valley Authority (TVA) Kingston Fossil Plant Public Health Assessment, from the Tennessee Department of Health (2010), recognizes that this incident is considered "one of the largest environmental disasters in U.S. history" (p. xvii). Additionally, the force of this spill caused the rupture of a gas line, and disrupted power and caused evacuation of affected neighborhoods (Tennessee Department of Health, 2010).

How does coal which is being processed inside of a fossil fuel plant become a physical hazard to surrounding communities? When the coal is burned, a waste product is produced in the form of a wet sludge, or wet coal ash. In disposing of this wet coal ash, the coal plant pumps the waste product to separate holding areas, or impoundments. Once the coal ash drains, it is removed and stored in an ash pond, and contained by dikes made from bottom ash. The coal ash is contained in these holding areas, in order to maintain a safe separation between the waste and the general environment and human population. At the Kingston facility, one of the retention walls, a dike, surrounding the wet ash ponds failed, causing the hazardous coal ash spill.

The danger to local communities is that coal ash has various toxic elements, including arsenic, mercury and polycyclic aromatic hydrocarbons (PAHs). These toxins remain even after the burning of the coal, and a spill causing the waste to escape, can be hazardous to nearby communities and the environment. If the wet ash dries to a degree that it becomes airborne, the toxic elements can be inhaled causing health problems of the upper airway, and lungs and "especially for those persons with pre-existing respiratory or heart conditions" (Tennessee

Department of Health, p. xxxi, 2010). Following the spill, the Environmental Protection Agency (EPA) (2012), provided oversight during water sampling and air monitoring to identify threats to human health. The Initial data found "samples of untreated river water had elevated levels of suspended ash and heavy metals known to be associated with coal ash", "and, (this) was observed again after a heavy rainfall" (EPA, 2012, p. 1). The drinking water that was sampled had been found safe to drink, so the primary health threats from this spill were the contaminated water in the river, and the threat of airborne coal ash. Secondary concerns are undetermined long-term health issues. The fear of contamination and subsequent health problems stemming from this incident caused fear and alarm for residents along the rivers.

Landers (2009) stated that the TVA had inspected the containment areas quarterly for seepage, and each day the containment areas were inspected visually. Though the facility and containment areas holding the coal ash were routinely inspected, the TVA stated that there were no significant problems found, and no indications that the dike was going to fail. In August of 2012, a federal judge ruled for over 800 plaintiffs, that the TVA was liable for the incident, due to negligence in the design of the containment ponds.

The risk that these containment walls, holding the coal ash, could fail was a major oversight on the part of the TVA. The risk that was not given proper attention, nor mitigated, was that a failure would cause such an enormous disaster to the surrounding community and environment. The fact that inspectors did not realize any type of instability in the containment areas indicates that quality assurance standards and/or skill levels could be an issue. Coal ash, is a by-product of coal plants all across the U.S., and this incident at the Kingston facility is an indication of how coal ash spills contaminate, and can be a serious health and environmental issue. The U.S. Energy Information Administration (EIA) (2013), states that there are 6,600 coal-fired power plants currently producing electricity in the U.S. The fact that even with regular inspections, this coal ash containment wall failed in the Kingston facility supports the argument that fossil fuel related accidents are not able to be controlled to acceptable (or non-damaging) levels. In this case, not only were homes severely damaged by the coal ash waste, but land and water were rendered unusable and contaminated by waste toxins.

Fossil fuel pipelines are another example of structures which fail, rupture and leak causing detrimental results to the environment and the co-located communities. In June of 1999 in Bellingham, Washington a severe rupture in an Olympic company gasoline pipeline killed three young people, destroyed one home and devastated the Whatcom Creek, its tributaries and Whatcom Park area, including the destruction of fish and wildlife. The fisheries of these creeks were closed for weeks after the incident. The NOAA Office of Response and Restoration *Case: Whatcom Creek, WA* (2013), states that 236,000 gallons of the gasoline gushed into Hanna Creek and then into Whatcom Creek. This is a coastal stream running almost four miles through the city park, neighborhoods and urban industrial areas before eventually emptying into Bellingham Bay. As the gasoline traveled down river, the fumes were ignited causing the deaths of the young men, and causing severe damage to the natural habitats along the creek.

The Seattle Times (Brunner, 1999) recorded the devastation of this incident, particularly to the fisheries and the severity of destruction of the local salmon spawning grounds. The Restoration Plan and Environmental Assessment (NOAA et al. 2002), for the Whatcom Creek Gasoline spill, provides details of the impacts of this incident, pointing out the cultural and economic importance of the creeks, the park and surrounding area for human uses, and for the habitats supporting the wildlife and fish. The ruptured pipeline and subsequent fire along the water and forest area had a devastating impact to the Hanna and Whatcom Creek watersheds, and co-located residential and commercial development areas, as well as the water supply which serves the residents in the City of Bellingham. The Assessment reports almost a complete elimination of the aquatic biota which included over 100,000 dead fish, crayfish and amphibians, as well as dead beavers, otters, birds and small mammals (NOAA, 2002).

It is not inconceivable that this incident could have been much worse if more people had been fishing, swimming or boating in the affected area, or that the fire could have spread quickly to housing and commercial buildings along the area. Fuel ignition is a horrifying reality for these spills, especially when the fuel and fire has a pathway through rivers and streams. This type of tragedy does not just contaminate the area, but can immediately end the lives of humans and destroy large populations of wildlife and live habitat.

On July 26th 2010, an estimated 819,000 gallons of crude oil leaked from a ruptured Enbridge company pipeline into a tributary creek of the Kalamazoo River in Marshall, Michigan. The NOAA's Enbridge Case Assessment (2013) reports that the oil travelled approximately 40 miles down the river affecting bordering wetlands, forests, residences, farmland and commercial property. This was a significant spill of crude, and the NOAA's estimate appears to be lower than what was reported by the Environmental Protection Agency (EPA) (2012) which reports that oil response workers collected over 1.1 million gallons of oil and over 200,000 cubic yards of contaminated sediment from the River.

According to the Calhoun County Michigan Public Health Department (2013) the oil spill caused serious concerns regarding contamination of the water, air quality, and the direct contact with humans. The Center for Disease Control (CDC) (2011) classifies Benzene as a human carcinogen and a natural part of crude oil and gasoline. The Oil Spill Intelligence Report, *Enbridge Spill Health Effects Surveyed* (Hyder , 2011), states that three days after the spill, 61 homes were evacuated due to unsafe levels of benzene found in the air quality tests that were conducted. There were 320 people who live or work near the spill site that experienced symptoms including headaches, breathing problems, nausea and vomiting. In the Spring/Summer of 2012 a portion of the River was re-opened, while further assessment was required prior to re-opening the remaining river. Public health workers reported 145 spill-related medical visits three weeks after the spill. The importance and meaning derived from the Enbridge incident, is how devastating this fossil fuel spill was to the river and tributaries of the Kalamazoo River, and the resulting health problems. Specifically, the Benzene-toxic air found after testing, shows the dangerous health risks that affect these communities, in addition to the physical contact with the contaminated water and land.

    Fuel pipelines are static structures which provide some element of control, and ability to be inspected in a stationary configuration. However, petroleum drilling rigs are platforms which are mobile, and are moored into place during the extraction of oil. These rigs can be moved from place to place in the sea, and along coastlines as scheduled for the work patterns, while maintaining fuel onboard. During the movement of a rig, control can be rendered unstable, or is subject to human error in operation. The NOAA's Office of Response and Restoration (OR&R) (2013), provides details of the Shell drilling rig *Kulluk* which ran aground off the coast of Kodiak Island, Alaska in January 2013. The *Kulluk*, a 266 foot long floating drill rig, had been drilling in the Beaufort Sea south of Seattle, Washington and was being towed when heavy seas caused the rig to break connection with the tow boats and be cast off uncontrolled in the sea. The rig was laden with approximately 140,000 gallons of diesel fuel, and though no spill occurred, the grounding caused a major disruption to the fishing and crabbing industry. If the

Kulluk would have ruptured, spilling the fuel, this incident would have escalated quickly into tragedy for Kodiak Island fishermen and businesses.

Once grounded in the Bay near Kodiak Island, the *Kulluk* had to be thoroughly assessed for damages and leakage by the Coast Guard and NOAA support coordinators. During the assessments, shipping and fishing vessel activity in the region was also disrupted, with local residents of Kiliuda Bay afraid of a potential spill. A spill of any degree would have been disastrous to the crabbing season. This incident had potential for a disastrous outcome for the coastal fishing areas, and sea life near and around Kodiak Island, and particularly for Kiliuda Bay. Additionally, the grounding of the rig threatened the habitat of the endangered Stellar sea lion and salmon streams, and the harvesting areas for razor clams.

The Alaska Dispatch (Anderson, 2013) reported that the Kodiak Island fishermen, caught in a holding pattern during the grounding, were anxious to continue the haul of an annual harvest of 660,000 pound-quota of Tanner crabs. The grounding of the diesel rig caused the coastal community a significant measure of distress and worry about the ability to harvest their annual catch of fish and Tanner crabs. While the hull and spill assessments were being conducted, the *Kulluk* was positioned for several weeks in a favorite crabbing area in the Bay, limiting access and contributing to further worry. Additionally, there was an increased amount of vessel traffic from the agencies conducting the hull assessments on the rig, which caused additional disruption to the locale (Anderson, 2013).

The NOAA (2013) reports that diesel is considered to be one of the most acutely toxic types of fuel, even in small spills of 500-5000 gallons. Direct contact with diesel spills can kill fish, invertebrates and seaweed, crabs and shellfish. Based on this assessment of small diesel spills, it is apparent that spillage of some or all of the 140,000 gallons (of diesel) from the *Kulluk* would have been severe to the overall seafood industry of the Kodiak Island region. The *Kulluk* rig grounding indicates the continued limitations we have in controlling the equipment and the vessels used in the petroleum industry. When the control of these fuel-laden structures is lost, the results are clear; petroleum is spilled and damage to the human population, fish, ecosystem, and wildlife is immediately incurred.

Merchant tankers represent yet another element of potential accidents with regard to fossil fuels impact to society. Though the tragedy of the *Exxon Valdez* provided a significant baseline to learn from, merchant tankers continue to be a major risk for spills along the coasts, and are particularly hazardous in-port the marine facilities. In November 2007, the merchant tanker *Cosco Busan* crashed into the San Francisco-Oakland Bay Bridge. The NOAA Office of Response and Restoration (OR&R) (2013) discusses that this incident caused one of the largest oil spills in the history of the Bay, with 53,000 gallons of fuel spilling into the waters. This spill caused immediate disruption to the Bay and parts of the outer coast affecting fishing, recreation and transit activities. In this case, the oil spread along the waters that surround the city of San Francisco, and washed ashore affecting 3,367 acres of beaches, rocky habitats, tidal pools, wetlands and lagoons. The recovery of this spill was assessed to vary between months to years depending on the area and habitat type of the Bay area.

In relationship to the economic impact of the fishing industry, the NOAA, OR&R (2013) stated that within two to three months of the spill, the annual schools of Pacific herring that swim into the Bay to find shallow spawning grounds were affected. The herring eggs collected in the areas of the spill were dead or deformed. Even after most of the oil was no longer evident, additional studies of spawning in the Bay found that toxic characteristics of the oil spill affected up to a quarter of herring spawning in 2008. The NOAA's Remedial/Injury Assessment (2013) stated that the human impact was estimated at 1,079,900 user-days being lost for recreational fishing, beach use, surfing etc. The Restoration Plan began in 2012, four years after the disaster and following the court decision approving a $44 million settlement with the responsible parties. The significance of this incident is that merchant vessels continue to be a focal point of tragedy in our ports and coastal areas, by causing major fuel spills, fires and damages. Unlike a stationary rig at sea, or a fuel pipeline that ruptures, these merchant vessels navigate directly into anchorages and berths in our cities. This close proximity of spills, to the ports and towns, impact the economics of the port facilities, the local businesses, and the health of the nearby human populations.

All of the fossil fuel related tragedies discussed here have elements in common, though they have differences in geographic location, type of disaster and time of occurrence. We see that fossil fuel disasters are not limited to one general area, coastal littoral, river, or type of cause. Wherever there are merchant ships, pipelines or other equipment used to support fossil fuels and toxic wastes, there is potential for damaging contamination and pollution. And, where these incidents take place we find significant harm to humans, wildlife and environment. The incidents in this chapter provide empirical examples of the many negative effects to human culture, and to the environment in which we share with the many species of wildlife. Unfortunately, the damage does not just affect our livelihood, traditions and health, but also causes the loss of human life in some cases.

These incidents continue to take place due to the lack of full control of containment, and the lack of appropriate indications and warnings. As with merchant ships and floating rigs, the control is surrendered to heavy storms, navigational breakdowns and human error. And, the inspections and safety measures in place for static equipment, such as in-ground rigs and pipelines, are not preventing these incidents from taking place (as evidenced by the destructive spills discussed in the Enbridge and Whatcom Creek disasters).

The *Deepwater Horizon* rig spill represents a critical eye-opening event, not only because of the negative impacts which have occurred, or are yet to be determined, but because the incident could have been even more damaging. This spill was an uncontained, seemingly limitless flow of oil into vast regional fishing zones and human coastal populations. And, even with today's technology and past lessons learned, this enormous and deadly incident still took place. Without the great efforts undertaken to cap the spill, it would have continued to leak millions of gallons of oil, further contaminating the waters, coastlines and inner-coastal marshes. Whereas a merchant ship has a limited payload of petroleum, oil wells and gasoline pipelines can continue to spill, for days, weeks or months, without effective stop-gap measures. In the case of the *Deepwater Horizon*, it was painstaking innovative human intervention which finally stopped the spill.

These incidents, depending on the location and severity, and the ability for human intervention to halt and restore the habitats, can have long-term impact to the societies affected. The National Wildlife Federation (2013), states that scientists are still assessing the impacts of the *Deepwater Horizon* oil spill, and the damage from the oil and chemicals used in the cleanup may not be known for years. An important timeline to recognize is that the Gulf oil spill took place at the peak of breeding season for many species of fish and wildlife, so the toxic elements in the waters may have severe long-lasting disruption to the aquatic food web, not yet understood.

In comparison, the impact of oil on Alaskan fish and wildlife from the *Exxon Valdez* oil spill was severe to the fishing industry, and subsistence of the Native Alaskans. The herring population collapse was not identified until four years after the incident. The National Wildlife Federation (2013) states that for the species that do survive the exposure to the oil, that one of the long-term effects may be loss of ability to fight off diseases. Though the Gulf fishing industry has endured the disaster (for now), the oil's long-term impacts to the fishing, crabbing and seafood overall may not be experienced for years with future negative impact to regional socioeconomic well-being.

The spills and fossil wastes, not only cause irreparable contamination and damage, but also cause immediate severe impacts to the societies involved. The negative impacts cause disruption to daily life, first by assaulting the environment and property, then subsequently these incidents can paralyze local economy and cause short and long term health problems. This book will continue to provide scope of the impacts of fossil fuel to society, by discussing such evolving technology as hydraulic fracturing used in natural gas extraction, and the rise of oil production in the U.S.

CHAPTER 3

The U.S. Natural Gas and Oil Boom: Abundant Resources in Shale Formations

There is an unprecedented change in the level of fossil fuel resources in the U.S. as a result of the significant increases of undiscovered, technically recoverable, natural gas and oil. These substantial increases in availability, and subsequent extraction and processing, are affecting regional socioeconomics and are introducing new environmental concerns. Specifically, these new "undiscovered" resources are resident in the shale formations (shale plays) within the known fossil fuel basins located throughout the U.S. The application of new technologies and improved geologic data are providing the opportunity to extract these fuels; fuels that were previously too far underground and hidden in the tight shale formations. These resources were unobtainable by conventional extraction methods. Now, the ability to extract these accessible fuel reserves is causing a mad rush in drilling and production. Also, this activity is stimulating a major demand for upgrades and expansion of current oil and natural gas facilities. But, what are the impacts to society in this major phenomenon of vast new resources of natural gas and oil? While there are positive economic factors, and domestic energy sustainment attributes, there are also major environmental and human impacts to be considered.

One key positive factor associated from these changes is the stimulation of employment and jobs to support the regional growth and production in drilling. There is also a subsequent key factor in the revitalization of economically troubled communities. However, the increase of available fuel and the growth in production also means that negative environmental impacts are occurring. Some of these impacts are directly related to a degraded quality of human life, as well as negative impacts to the land and wilderness. Individuals whose properties and homes were once serene and untouched by industry are now in the midst of drilling and pumping facilities, fuel trucks and disposal pools. Additionally these changes are raising new environmental concerns such as negative landscape alteration and contamination of drinking water in the regional aquifers. All of these impacts, positive and negative, are affecting the societies within the U.S. regions of the shale formations. What is the overall impact as we move into the future? Will the environmental and quality of life concerns be offset by the economic stimulation and regeneration of failing townships? This stimulation is particularly important for the people who have been hurting for employment opportunities due to the faltering U.S. economy.

To better understand the changes and impacts taking place, the following illustrations provide insight on the developments relating to increased fossil fuel resources. There are many fuel basin regions (also referred to as shale plays) in the U.S. As a primary example, the Bakken and Three Forks shale formation (North and South Dakota, and Montana) are experiencing phenomenal development in undiscovered resources and growth. Another primary region, with recent discoveries, is the Marcellus shale formation situated along the Appalachians. Although these two shale formations are most prominent due to more recent activity, there are many known fossil fuel basins throughout the U.S. All of these basins are benefiting from new available fuels found in the shale formations. Other basins include the Fayetteville Shale of Arkansas, the Woodford Shale of Oklahoma, and the Haynesville Shale of Miss and Louisiana. The advances in today's fuel extraction technology and geologic assessments, impacts them all.

The following graphic from the U.S. Energy Information Agency (EIA) (2011) provides the most recent display of the shale plays in the U.S. lower 48 states. The graphic also provides a visual idea of the large scale impacts on society across the nation. Each one of these regions is being affected by the increased extraction opportunities which impact the economy, the environment and human quality of life.

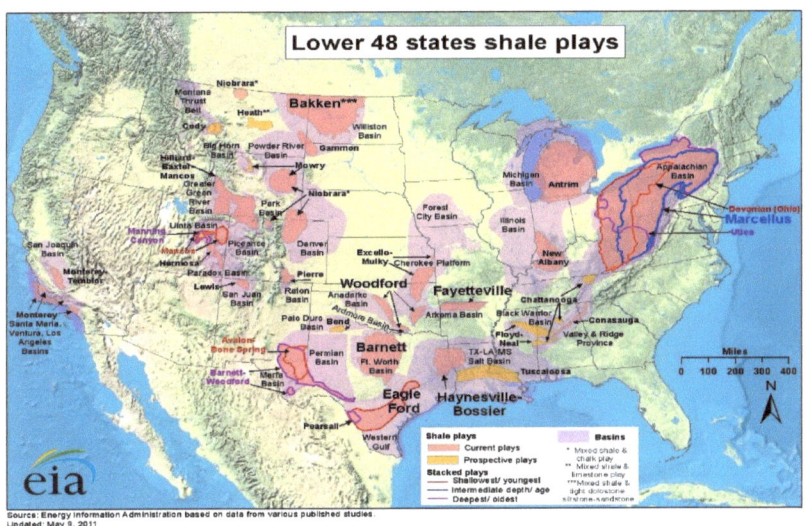

How has advanced technology in extraction, along with the new assessments of increased levels of fuel changed the impact in these shale plays? The advanced technology of hydraulic fracturing, or "fracking", and directional drilling are the key factors. Additionally, the continued scientific exploration and research has provided advanced geological assessments that support the uses of the fracking and directional drilling. Specifically, in the past two decades technology and science have been honed and applied to the conventional methods in the field. Current advances have opened up new and great possibilities in oil and gas extraction, and this has encouraged further exploration. The advances in horizontal drilling have provided a substantial access to the tight oil and tight gas located in the shale formations. Tight oil and tight gas are descriptive terms meaning the resource was practically unattainable in the recesses of the shale, far beneath the ground (compared to conventional access through vertical drilling). Fracking combined with the directional drilling is now allowing access to multiple points in the horizontal shale formations.

Science has played a very important role in supporting this drilling technology. Specifically, the U.S. Geological Survey Department (USGS) has been instrumental with providing new data studies. The information being introduced from new geologic assessments, of the known U.S. fossil fuel basins, has presented exceptional data targeting undiscovered and significantly higher levels of oil and natural gas. These new geologic assessments are providing immense opportunity for extracting these resources; resources expected to last for decades. Dobb (2013) states that the developing extraction technology is "in effect a skeleton key that can be used to open other fossil fuel treasure chests" (p. 1). The projections for these untapped fuel resource levels are immense. In the *Annual Energy Outlook 2015*, the Energy Information Administration (EIA) (2015) stated that production of natural gas in the U.S. increases from 24.4 trillion cubic feet in 2013, to 35.5 trillion cubic feet in 2040, which is a 45% production increase. Additionally, the domestic crude oil production increases as result of growth in tight oil production through 2022 where production is expected to reach 7.4 million bpd.

The importance in the substantial increases, specifically in that of the available natural gas reserves, is that the U.S. will have an abundant domestic supply (more than it consumes) for several decades. This abundant supply has also allowed the U.S. to increase domestic natural gas exports. The EIA reported that natural gas exports increased by 33 percent to 1,507 barrels per cubic foot in 2011. The geologic surveys and ongoing drilling exploration indicate increased levels of fuel availability in each one of the lower 48 fuel basins. The Williston Basin, comprising the Bakken and Three Forks shale formations (North and South Dakota, and Montana), is the most currently recognized shale region with significant assessments of undiscovered, recoverable oil and natural gas resources. The USGS (Charpentier, et al. 2013), reports an estimated undiscovered volume of 7.4 billion barrels of oil, 6.7 trillion cubic feet (Tcf) of natural gas and over a half billion barrels of natural gas liquids in both of these formations. This 2013 assessment is a significant increase; double that of the last estimation in 2008. The USGS (2011) also reported that the Marcellus formation (in the Appalachian Basin) has approximately 84 trillion cubic feet of technically recoverable natural gas, and 3.4 billion barrels of natural gas liquids, compared to the previous 2002 estimate indicating approximately two trillion cubic feet of gas and .01 billion barrels of natural gas liquids. Like the Bakken formation, this 2011 Marcellus reassessment also represents a significant increase from the previous Appalachian Basin estimates.

The new fracking and drilling technology has been incorporated into the current processes, allowing the access to these significant quantities of previously unobtainable deposits. Mooney (2011) explains hydraulic fracturing, which originated in conventional vertical drilling, and provides insight on accessing the shale reserves. During the conventional fracturing, contact is made with layers of shale during drilling. Chemically treated water and sand are used in extremely high pressures, blasting down the well into the shale. The hydraulic application is the extension methods of the drill as it is positioned further down in the ground, and along the shale formation. This process is causes cracks (or fracturing), giving way for the trapped natural gas to escape. The force of this pressure causes the gas, or oil, to be expelled back to the surface.

Today, the significant developments are how the drilling processes have been combined with directional and horizontal drilling substantially increasing access to the fuels. This directional ability and access allows the drilling to continue for thousands of feet parallel to the shale formations. This parallel drilling exposes significantly more gas and oil resources that are trapped in the shale. Mooney describes the significant results as "a veritable Gas Rush, sequestered layers of methane-rich shale have suddenly become accessible" (p. 1). The fracking methods provide the means to loosen these resources forcing it back to the surface, along with the chemically treated water. This new technology is also impacting the land and environment as more drilling facilities and equipment are required. Additionally, this process also incorporates the use and disposal of chemically treated water near fresh water resources.

Hamilton and Slonecker (2012), of the U.S. Geological Survey (USGS), report on geologic findings which show significant landscape changes associated with natural gas resource development. Their report specifically targets the counties of Bradford and Washington which are located in the northwest region of Pennsylvania. These counties are situated within the Marcellus shale formation. These land disturbances, such as well pads, roads and pipeline construction were observed and studied from high resolution geospatial imaging. In Hamilton and Slonecker's report, the Director, USGS, Marcia McNutt, states that "the widespread use of hydraulic fracturing to produce natural gas and coal-bed methane in these counties has unlocked new sources of energy, but it is also modifying the landscape at an unprecedented rate compared with other forms of energy development" (p. 1). As the landscape disturbances are realized, such as large scale loss of forests, there can be significant impacts to ecological resources.

The assessments from the USGS provide empirical data and research which can help assess the impacts of land disturbances from fracking; impacts such as alteration of water quality, wildlife and socioeconomics of the regions. For perspective, as example, in the Marcellus Shale development the USGS study reports that in the Bradford County area, there are 642 gas sites that have resulted in disturbance of 1500 hectares. These disturbances include 45 miles of new roads, and 110 miles of new pipelines. New roads and transportation access development are common land disturbances and identifying characteristics for all the fossil fuel basins in the U.S.

Another environmental concern of fracking is the large amounts of fresh water needed to conduct the operation. This water has to come from local sources in many cases. The same sources also support the regional populations with fresh water supplies. The use of fresh water for fracking operations is a detriment to regions that are experiencing droughts, and are already in short supply. The EPA (2012) is conducting water shortage studies in the Susquehanna River Basin and Bradford County in Pennsylvania, as well as for the counties surrounding the Colorado River Basin. The amount of freshwater used in fracking (average of 4 million gallons per surface well) is a significant amount to source from local freshwater supplies. According to the EPA (2012), this requirement of fresh water for fracking, along with increased populations due to production, creates a significant demand on freshwater sources.

Another concern is that spills and leakage of contaminated liquids that fracking produces, can contaminate the local fresh water supplies. Once brought back to the surface, the fracking water is stored in pits or ponds near the extraction sites, or it is re-introduced into the ground through wells for holding. This liquid waste, if not contained, can find its way into the underground fresh water supplies. Examples of cases involving uncontained fracking waste fluids are provided in the 2012 EPA study. One specific case is the examination of water resource impacts from a well "blowout" in Killdeer North Dakota (ND), within the Bakken Shale region (p. 137). Killdeer is a rural county with a population of 3500. This case involves fracking fluids which broke from containment when an intermediate well casing burst. The cleanup involved removal of 105,252 gallons of fracking fluid, and 2,000 tons of soil that had been contaminated by the fracking fluids. The Killdeer Aquifer is located under the study site. The EPA study is pending.

In a separate study, Proceedings of the National Academy of Sciences (PNAS) reported that methane had been detected in drinking water wells within the Marcellus shale region of Pennsylvania, where fracking is taking place. The study (Darrah et. al., 2013) reports an 82 percent detection of methane, from the 141 drinking wells tested across the northeastern Appalachian region of Pennsylvania. The report indicated that drinking wells located nearer to the fracking activity had a higher concentration of methane, six times higher for homes within one Kilometer from the gas wells. The studies of fracking impacts to local fresh water supplies are still being conducted, but the data from the EPA, and the PNAS indicate that fracking does pose spill-over contamination risks, especially for areas and homes nearer the gas wells.

The EPA's Progress Report on the Study of Potential Impacts of Hydraulic Fracturing on Drinking Water Resources (EPA, 2012), provides current data and detail on the research being conducted on the impacts caused by fracking. The EPA states that there are concerns associated with natural gas and shale gas extraction, and fracking, and that the operations cause impacts to the environment. The EPA's Progress Report has identified over 1000 chemicals found in the flow-back water used in fracking; however the results of the exposure of the drilling water to drinking water resources have not been finalized. The graphic depicted below provides a view of the high pressure well injections that occur during fracking (both vertical and horizontal). The graphic indicates how the fracking wells penetrate or are in close proximity to fresh water wells and resources. When a well casing ruptures or fails, or when fracking fluids leak from the well (underground, or at the injection point on the ground), then there is potential for fresh water contamination by the flow-back water and chemicals, as well as petroleum. The EPA's study of well injections will provide data on changes in local drinking water reserves, by looking at construction failures in wells such as in the vertical or horizontal phases.

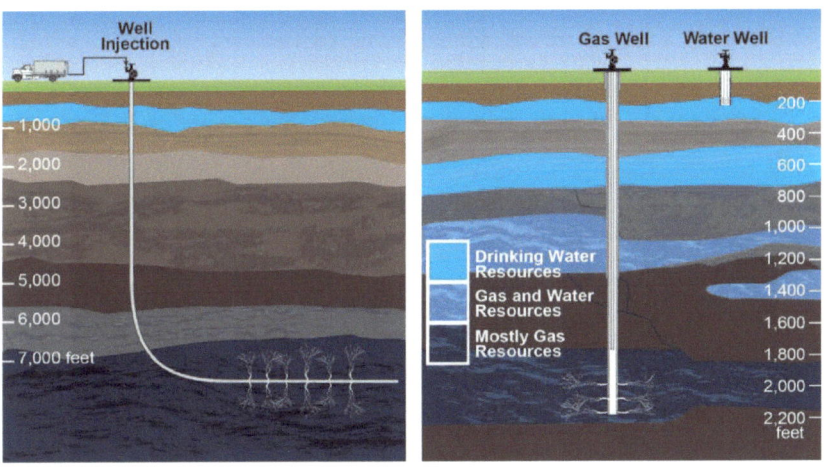

(EPA's Progress Report on the Study of the Potential Impacts of Hydraulic Fracturing on Drinking Water Resources, Section 2, page 17, 2012).

In addition to potential water resources contamination, fracking processes also affect the human quality of life when homes and communities are surrounded by the oil and gas production facilities and equipment. The concerns include fumes/air emissions from the facilities and holding ponds, noise pollution, and the clearing of forest and land areas for the facilities and wastes. Lavelle (2010) describes family life in regions that are experiencing the gas and oil boom, specific to quality of life in what was once a scenic and perfect environment. Lavelle discusses the impacts to quality of life of the Hallowich family, in the Mount Pleasant Township, southwest of Pittsburg, PA, and how they have experienced the detriment of a natural gas extraction and processing around their home.

The Hallowich family searched for their property and found it in a wooded area where they built their home. Unfortunately, their property is situated on one of the most active parts of the Marcellus shale formation. Their beautiful views and serene country-side habitat eventually became occupied by natural gas wells, processing equipment, and underground pipelines. There was also a three acre waste holding pond (plastic lined) co-located in the area, and gravel roads which were developed around their property to provide tractor-trailers access. The quiet and natural environment of their 10 acres had been overtaken by sounds of trucking as well as smells from the diesel fumes from the fracking process.

The Hallowich' family are victims of what so many other families have discovered from living in the areas of the natural gas and oil industry. The reality usually comes as a surprise when land owners discover that they do not own the mineral rights under their property. In these cases when the mineral rights are sold by previous owners of the land, oil and gas companies can drill horizontally under the properties. The Hallowich family receives some compensation, of a few hundred dollars a month, but this does not cover what the family has lost in quality of life and the fresh water that they have purchased and delivered. The family claims that the waste water from drilling contaminated their fresh-water well, so they have not used their water since, for fear of the health risks. The fumes that come from the petroleum, of the oil and gas pumps, and from the holding ponds are part of everyday life now.

The negative aspects to the society and the environment are significant, however there are also positive changes taking place also due to this boom in gas and oil production. There is socioeconomic significance in these discoveries particularly in the thousands of new job opportunities becoming available. These jobs are essential in supporting onsite production and the trucking requirements. As well, jobs are needed in development of the supporting infrastructures of these regions, specifically in the building or improving roadways, railways and addressing new power requirements. In some of these areas, particularly the Bakken Shale formation, the natural gas boom is reviving towns and rural communities that were on the edge of collapse due to economics and declining populations. As well, due to the influx of workers, populations are on the increase in many regions. So, once-destitute communities have promise for survival as the economy is strengthened by the growth taking place.

Boyd (2012), a correspondent for the American Oil and Gas industry, provides an industry perspective of the increased oil production seen in the Bakken shale formation. Boyd states that the number of wells expanded from 470 to 3337 between 2008 and 2012. The author reports on the numerous challenges that both the operators and municipalities are facing with the demands required to support this (gas and oil) boom. Boyd states that "operators and municipalities find themselves wrestling with infrastructure challenges many regions would love to face: crude oil transportation constraints brought on by surging production, and maxed-out municipal services in a rural region with a rapidly expanding population that is bursting at the seams" (p.1). Boyd indicates that in the North Dakota region of the Bakken shale basin, the demand is significant, with requirements for drilling and pumping facility equipment, housing and office facilities.

Dobb (2013) provides vivid on-site descriptions of the Gas and Oil boom evolution in North Dakota. Dobb surveys the changes happening to the landscape, and provides empirical perspective through interviews of local citizens and the individuals trekking into the region looking for work. North Dakota comprises a major region of the rich oil and gas reserves available in the huge Bakken shale formation. This North Dakota region is described as rural and barren, with little supporting infrastructure such as adequate roads, and local businesses that are required for providing supplies and materials. The author observes that this lack of infrastructure has created the need for workers who can help build the roadways, provide logistics, and set up the production sites. As well, there is an increasing demand specifically for experienced truckers who can transport oil, gas and wastes out of the area, and transport needed supplies and fresh water into the production sites.

As Dobb observes, this rural environment, the prairie, is quickly being developed for major industrial use, and individuals who are looking for work are descending upon the northern region of Bakken in thousands. Labor positions, such as oil rig workers are providing locals with jobs, however in many cases more-advanced positions are filled by experienced rig workers transferred from other oil operations areas. So the employment opportunities are not only supporting the local populations in the Bakken region, but also are requiring a number of experienced workers to transfer from other parts of the U.S. The author describes many of the transient workers as mostly men, who hold up in temporary camps made up of mobile homes, cabins and RVs. Towns hope to establish permanent living communities for these workers and their families, eventually developing the social infrastructures such as housing, schools and hospitals which are needed to support this growth.

According to information provided by the North Dakota Department of Commerce (2013), the U.S. Census Bureau reported that North Dakota had areas ranking as "some of the fastest growing in the nation" (p.1). The Census Bureau reports annual population increases including gains in the following micro (populations of at least 10,000) and metro (populations of at least 50,000) areas between July 2011 and July 2012:

- Williston Micro area gained 2,281 (9.3% growth)

- Dickinson Micro area gained 1,624 (6.5% increase)

- Minot area Micro area gained 862 (1.2% increase)

- Bismarck Metro increased in population (2,776 residents)

- Fargo Metro gained (3,726 residents)

- Grand Forks Metro gained (827 residents)

Rod Backman (chairman of the North Dakota Census Committee) confirmed that the state is attracting new residents who are seeking the growing employment opportunities. The North Dakota Department of Commerce states the population increases are due to the rapid growth in the oil industry. Additionally the Department of Commerce acknowledges that the state is supporting "historic funding for infrastructure improvements and enhancements for public health and safety, affordable housing, child care and emergency services" (p.1).

The social impact of this "gas and oil boom" is the major change being experienced in the towns and communities, especially in the farming and rural communities. What once were small communities are now becoming large hubs for individuals seeking homes and food while working for the natural gas and oil sites. These rural landscapes are being covered with pipelines, extraction wells and fuel storage sites. But, along with those industrial structures, come opportunity for much needed jobs, and hope and security for workers and their families plagued by economic distress. However, Court, Jackson & White (2012) discuss important concerns for local economic development taking place in the rural areas and small communities of the shale regions. How will these communities, particularly policymakers, meet the needs of both citizens and the industry during this extensive growth and change? The planning and decisions will have to consider how to sustain and balance the economic diversity, and future development of the region. The development and understanding of policies will be a significant undertaking, and participation of all stakeholders is an essential element.

Dobb (2013) also discusses the "human" aspect of the gas and oil boom taking place in the Bakken Shale formation, providing a lens in which to view family security, and the harsh work-life that workers are enduring in order to make a living. Dobb follows a 39 year old mother, Susan Connell, through the treacherous workday that she faces as a self-taught driver of an 18-wheeler. Connell is one of the few women working the laborious jobs associated with this industry, but it is what she has to do in order to pay the mortgage, and to help support her husband and two young daughters. This young mother leaves her family behind to make the seven hour trip from Montana to North Dakota in order to travel to the job sites in North Dakota. Dobb states fracking has increased the demand for jobs, especially for tuckers to haul dirty water and waste fluids which are created during pumping of oil and gas. Dobb describes the laborious process that Susan Connell experiences during her waste disposal trips that consist of climbing up the ladders to the holding tanks in order to measure and siphon out the dirty water with large heavy fire hoses.

The fumes from the oil and fuel in the holding tanks are overwhelming, and, the fumes from the water waste which contains numerous chemicals such as hydrogen sulfide are horrendous, especially when the hatches of the holding tanks are opened. Connell described that the first time she conducted this process, she fell to her knees from the overwhelming toxic fumes coming out of the waste water holding tanks. One time she was pumping waste water, and was overcome with breathing the fumes and suffered subsequent stabbing pains in her stomach and vomited for a week (Dobb, p. 1, para 2). This example of harsh work-life for the individuals supporting the shale formation fuel growth and production is common. The characteristics are the same in the dynamic expansions of this industry includes long hours and exhaustive physical labor. Importantly the health of the workers is a great concern, where they often are inhaling fumes and come in contact with fuel and the chemically treated water from the drilling. While the demand for workers is providing jobs in this industry, the conditions of the work are a negative consequence in many cases.

There are important and substantial changes taking place in the evolving shale formation fuel extraction trend. These developments and production growth are causing changes, both positive and negative, to societies and the environment particularly within these fuel basin regions. The development of advanced technology in hydraulic fracturing, and new assessments from geologic surveys are key factors in this gas and oil "boom". The fuel production is supporting the rejuvenation of decaying communities, and stimulating the influx of income to farmers and landowners who have been battling economic downfalls. However, for others in these same areas, the oil rigs, and natural gas pumps, and waste from extraction of drilling is degrading their quality of life, and negatively impacting the environment.

How we foster in this evolution of new available resources, and how we make decisions on policy and direction for this growth is extremely important, and not an easy task. This is a critical era which impacts local and national economics, and the protection of our environmental resources, wildlife and landscape preservation. Blake (2103), from the U.S. Department of the Interior, provides discussion on these vast new fuel resource potentials and shares a very fitting quote from the Secretary of the Interior Sally Jewell, "we must develop our domestic energy resources armed with the best available science, and this unbiased, objective information will help private, nonprofit and government decision makers at all levels make informed decisions about the responsible development of these resources" (p. 1). Secretary Jewell's statement provides an intellectual certainty, and fortifies the fact that a successful energy future is dependent on solid leadership and the inclusion of scientific voice.

CHAPTER 4

The Future of U.S. Energy: Leadership, Alternative Solutions and Collaborative Strategy

In order to address the U.S. requirements for sustainable energy in the future, and to decrease the dependency on fossil fuels, it is imperative that alternative energy technologies continue to be researched and improved upon. Additionally, it is essential that leadership at national and local levels set realistic agenda and make collaborative decisions. Realistically, society will continue to rely on fossil fuels to some extent for decades, particularly in the oil we use for transportation; however the negative impacts of this dependency demand continued research and application of alternative energies. The primary stakeholders of science and government must provide collaborative, effective contributions in forming the nation's energy direction. The U.S. government agenda plays a major part in much of the energy related decisions and strategies. The agenda will set the tone of leadership and represent the direction that the country is heading by defining the resources needed to conduct research and fund projects. As well, clear agenda and strong leadership provides focus and supports public awareness. If this is true, then what direction is the U.S. government heading with regards to the nation's energy goals? Is our national strategy indicating progress? And, in the overall strategy is the U.S. participating in global efforts for common goals in alternative energy solutions?

A well-defined strategy is essential for implementing the research and resources for the development of alternative energy. Key stakeholders need to also provide clear communication for public awareness, and demonstrate strong leadership decisions. Clear communication will provide focus on the value of balancing scientific research and presidential agenda. Currently there are numerous promising initiatives and experimental projects underway with alternative energy; but are the primary stakeholders working together to form the best strategies for achievement? Is the technology put in place today working, and is research focused on smart and realistic energy solutions? The combined efforts of all-concerned (government, industry and science) can help move our society away from the current level of reliance we have on fossil fuels. Additionally, it is important that the U.S. agenda and the U.S. scientific community include collaboration as a key strategy for energy planning. As well, the U.S. strategy should include international participation in energy research and initiatives, particularly because of the common interests in combating global warming and fossil fuel dependency.

President Obama's U.S. Energy strategy states three distinct pillars of action which are focused on 1) clean energy economy 2) issues of climate change and 3) protection of the environment (Whitehouse, 2013). The administration encompasses an "all-of-the-above" approach in answering the nation's energy requirements. Balanced with alternative energy goals, this "all-of-the-above" agenda includes continued drilling and exploration of gas and oil (though with the caveat of safe and responsible production). The Whitehouse Blueprint for a Secure Energy Future (2011) discusses that domestic petroleum resources, particularly oil and natural gas, are significant components to national energy. As well the Blueprint points out that the development of these domestic fossil fuels is essential and "enhances our energy security and fuels our Nation's economy" (p. 9). This statement indicates the acknowledgment of the importance of domestic petroleum, at least for the next few decades, and highlights the positive aspects of domestic fuel by noting that domestic oil production is at the highest level since 2003.

However, (here's a key balancing point to the fossil fuel resource issue), the President's Blueprint specifically recalls the *Deepwater Horizon* tragedy, and draws specific attention to the need for safety reforms, and the environmental standards for oil and gas exploration. The safety reforms in the President's agenda are implemented through improved oversight by the

Department of the Interior. Additionally, the President is holding industry responsible for the safe and effective production of these resources, and encouraging development of new safe uses for the nation's abundant natural gas reserves. It is apparent that the President's agenda is attempting to carefully balance the numerous variables associated with the dependency on fossil fuels. Specifically, by acknowledging the advantage of utilizing the vast domestic resources, while at the same time attempting to provide assurance that further tragedies like *Deepwater Horizon* will be kept in check through "oversight".

The President's agenda is basically nurturing a tight-rope plan where on one side are the perils of continued fossil fuel uses (impact to human and environment and reliance on foreign imports), and on the other side is the research and technology that will move us into a less fossil fuel dependent society. The balancing rod atop this "tight-rope" is in maintaining an effective oversight on the human and environmental safety issues, and developing and tracking smart energy strategies, while continuing to move toward an effective alternative energy future. This means that the President's agenda for the nation's energy direction meets with challenges of great opposition on both sides of the key issues, particularly between the environmental supporters and the petroleum industry leads. However, this agenda does find balance and does demonstrate defined approaches for progress amongst the obvious conflicting issues associated with fossil fuel requirements. While the President cannot ignore the new found resources of the shale basins (or the already available domestic sources of fuel), the government's oversight and reforms at least provide an effective solution to address the negative impacts.

The important factor in this agenda is the President's strong and continued focus on alternative energies. And, though reforms and oversight by the Department of Interior indicates progress (or control to some level), this still does not provide a long-term solution for protecting our way of life, or the air and environment. The progress and growth in alternative energy technologies is the important conduit which will allow society to fully cut the umbilical cord that binds us to reliance on fossil fuels. Even so, is alternative energy research and application being fully exploited? Is there an effective collaboration between government, science and industry that will facilitate a successful transition from fossil fuel dependency to alternative energy solutions? And, is the U.S. taking advantage of international collaboration efforts that will strengthen or multiply the technological gains achieved by shared teamwork?

Alternative energy technologies and research are making valuable headway, and the U.S. is involved in numerous collaborative agreements and projects. The U.S. government sponsors many energy initiatives under the auspices of the U.S. Energy Department, the U.S. Environmental Protection Agency, and through grants provided to scholastic institutions. Today's technologies in alternative energy solutions are numerous, and the direction for research and application of those must focus on common goals, particularly as funding is a primary concern. In the case of cooperative energy research and science, information and technological development is shared by industrial partners within the U.S., and also internationally. The collective benefits from many expert sources can be enormous as personnel, hardware and software, technology and collected data can be shared for a common goal. Specifically, the U.S. has significant energy related involvement in international forums and projects through the U.S. Department of Energy's (DOEs) Office of Fossil Energy.

One multilateral organization is the Carbon Sequestration Leadership Forum (CSLF). The CSLF is an international forum, of 22 countries and the European Commission. The CSLF (2013) provides insight to the mission and states that the organization "is a ministerial-level international climate-change initiative that is focused on the development of improved cost-effective technologies for the separation and capture of carbon dioxide (CO2) for its transport and long-term safe storage" (p.1). An example of a recent major CSLF project is the Air Products CO2 Capture from Hydrogen Facility Project. This project, currently under development in Texas, was nominated originally by the U.S., United Kingdom and the Netherlands. The CSLF (2012) describes this project as a large-scale demonstration which intends to prove the ability to capture and purify Carbon Dioxide (CO2) for sequestration. In the stated goal, this project is expected to recover approximately one million tons per year of C02, from the designated hydrogen facility in Port Author, Texas. Once captured, the CO2 will be transported along pipelines where it will be then be utilized in Enhanced Oil Recovery (EOR).

The status of this project appears to have had a positive start. Air Products (2013) reported the achievement of a key milestone of the CO2 sequestration through a successful demonstration of the operation. The significance of this project is that approximately 75% of CO2 produced at a hydrogen facility will be captured instead of being released into the atmosphere. For the U.S., measurable and effective decreases of CO2 emissions is a significant goal to achieve; but along with CO2 sequestration projects, other clean-energy technology also provides positive gains which helps satisfy safe energy objectives. For example, the U.S. is actively involved with collaborative energy research projects being conducted with nuclear fusion. LePoire (2011), an environmental analyst at Argonne National Laboratory, discusses alternative energy sources and international collaborative efforts, particularly in nuclear experimental research. LePoire discusses the International Thermonuclear Experimental Reactor project (ITER) which represents a major scientific endeavor to produce energy from nuclear fusion. Through international cooperation and scientific exchange, ITER seeks to accomplish a commercial fusion solution that will offer the world energy with no carbon emissions, air pollution and provide unlimited fuel. The ITER project recognizes that fossil fuels were significantly damaging the environment through Carbon Dioxide emissions and greenhouse effect gases. The ITER is represented by seven international teams including the U.S., China, European Union, India, Japan, Korea and Russia.

A major benefit of the ITER collaborative research efforts is that national representatives can share in the overall data pool derived during technological and scientific development. Each contributing member also enjoys the limited costs benefits of the collaborative design, compared to that of self-investment of time and funding. For example, The U.S. ITER (2012) states that U.S. cost for construction in the ITER projects is less than 10%. The U.S. ITER branch is part of the Department of Energy (DOE) Office of Science project. The U.S. ITER research team is comprised of scientists and engineers from over 300 companies and universities, including the Princeton Plasma Physics Laboratory and Savannah River National Laboratory. The U.S. ITER (2012) describes ITER as an "unprecedented international collaboration of scientists and engineers" with the mission to demonstrate the "scientific and technological feasibility of fusion power for the commercial power grid" (p. 1). Specifically ITER is a large-scale project, a major experiment for demonstrating the possibility of capturing fusion energy for commercial testing using key technologies.

The U.S. is also involved in bilateral cooperative initiatives with 13 countries, supporting efforts in fossil energy research and clean energy development. The U.S. participation in these bilateral clean-energy projects is not just a self-serving obligation; the exchanges offer numerous winning benefits to each member of the collaborative team. To provide an example, these bilateral agreements include joint initiatives with India and with China which efforts promote effective use and production of coal, oil and natural gas. The U.S.-India Energy Dialogue is a successful cooperative exchange that provides focus through energy resource specific working groups (coal, oil and gas, nuclear, power grids, etc.). The U.S. State Department (2013) provides insight on keynote discussions from the June 2013 U.S.-India Strategic Dialogue conference held in New Delhi. U.S. Secretary of State John Kerry attended this dialogue, which was hosted by the Indian Minister of External Affairs, Salman Khurshid. Minister Khurshid reaffirmed the collaborative efforts of both the U.S. and India to "ensure energy security, combat global climate change and support the development of low-carbon economies" (p. 1). Secretary Kerry and Minister Khurshid both acknowledged that the benefits from this continuing Strategic Dialogue will support the creation of many opportunities and jobs in both countries.

A recent example of the ongoing technical collaboration between U.S. and India is shown in the oil refinery sector. In May 2013, representatives of Indian state-owned refineries visited with U.S. experts that specialize in refinery efficiency. The visit offered technological exchange for Indian refiners in learning techniques and applications for improving the oil processing through slurry-hydro-cracking. This "slurry-hydro-cracking" process allows more efficient conversion of residue in oil refineries. In addition to technological exchange and team building, this cooperation subsequently presents further opportunities for U.S. firms that are involved in this exchange.

The U.S. and China, the two largest producers and consumers of energy, have worked in various diplomatic cooperative scientific and technological efforts since 1979. The cooperative efforts are continuing today and expanding into clean-energy issues. The U.S. Department of Energy (2011) reported that in June 2008 the U.S. and China established a new phase of partnership against emerging challenges of global energy security and global climate change. This partnership created the U.S.-China Cooperation on Energy and the Environment, which included cooperative plans of action for "energy efficiency, electricity, transportation, air, water, wetlands, nature reserves and protective areas" (p.1). In 2009, during the Beijing Summit, President Obama and China's President Jintao introduced additional U.S.-China clean energy initiatives. These initiatives further established energy cooperative objectives where U.S. and Chinese Scientists are working together in both countries. These U.S.-China scientists are partnered in finding solutions on key issues such as development of clean coal and carbon capture technology. As well, these U.S.-China teams are developing electric vehicle and battery technology and working together on the acceleration of access to shale gas resources in China. U.S. Department of Energy, Secretary Steven Chu states that cooperation between the U.S. and China "can greatly accelerate progress on clean energy technologies" that is beneficial for both countries (p. 2).

Recent leadership representation continues to show the importance in the U.S.-China exchanges. In April 2013, Secretary of State John Kerry addressed the U.S.-China Energy Cooperation Event in Beijing, reaffirming the importance of U.S. and China action and cooperative efforts in combating the 21st century energy and environmental challenges. Secretary Kerry acknowledged progress in the U.S.-China Energy Cooperation Program such as the aviation biofuel project, where in 2011 a successful demonstration flight took place in China with a jet using biofuel. This demonstrative accomplishment was a result of U.S. aviation companies such as Boeing and Pratt, with Chinese organizations in developing biofuel from biomass products grown in China. This project will continue the research and development in support of sustainable flight through the use of biofuel. This type of substantial progress can have significant future impact on the decrease or replacement of diesel fuels in aviation.

Additionally the U.S. and United Kingdom are teamed together under a November 2000 Memorandum of Understanding (MOU). This U.S.-UK MOU provides agreement for collaborative research and development toward environmental protection and energy security and exploration of the "opportunities for expanded fossil energy utilization" (p.1). Specifically the MOU provides guidance in the fields of fossil energy, renewable energy, waste-related management and environment, and other energy related technologies and systems (Office of Fossil Energy, 2013). Currently, the Office of Fossil Fuels (2013) states that the UK-US collaboration is involved in development of advanced high-temperature materials for advanced fossil power plants. Also, the U.S. Air Force and UK's Royal Air Force have a cooperative relationship for reduction of fossil fuel use in aircraft, and are focusing on alternative fuel solutions.

It is not inconceivable that these bilateral agreements of technological exchange will "spill over" into a global shared knowledge and application, beneficial for the specific issues in each country. These agreements have particular emphasis in research involving clean energy technology, and experiments in carbon sequestration. These initiatives have a significant and important global impact to reduction on fossil fuels use, and a cleaner environment.

Collaborative initiatives are also prominent with the U.S. between industry, government and scholastic institutions, and importantly are bolstered by innovation. The National Renewable Energy Laboratory (NREL) (2013), under the Department of Energy (DOE), has reported a recent collaborative initiative teaming with the U.S. Navy, Cobalt Technology and Show Me Energy Cooperative, to find an economic solution to domestic transportation fuels. This current effort is based on the conversion of renewable biomass materials (from switchgrass) into butanol, then into further production of jet fuels. The goal for this initiative is to demonstrate a cost-effective alternative for navy jet fuel. Muller (2012) states that among all the biofuel initiatives, that switchgrass have rapid growing characteristics, and is the best choice for a "significant biofuel component to solving our energy problems" (p. 219). The U.S. Navy utilizes an enormous amount of jet fuel during exercises, logistics and in operations, so a successful achievement from this research has significant possibilities, particularly in cost savings and environmental aspects. An alternative fuel source, particularly from biomass products could significantly decrease the Department of Defense (DoD) dependency on petroleum based fuels. NREL has already demonstrated the ability to convert biomass to butanol, according to a Senior Project Leader, through successful test runs in 2012. The NREL (2013) states that a benefit from successes from this project is the production of jet fuel in the U.S., stimulating domestic jobs. Additionally, success of this project would result in positive impact to the U.S. energy security, and also reduce greenhouse emissions significantly (by 95%) in comparison to the emissions of today's jet fuel use.

  A successful U.S. energy plan for the future is dependent on focused leadership, advancing our alternative energy technologies, and embracing collaborative strategies. It is essential that the President's energy agenda be balanced with smart and safe uses of the domestic fossil fuel resources, and the aggressive research and application of alternative energies. Additionally, the U.S. energy initiatives will be significantly strengthened by innovation and collaborative strategies which are being conducted today through participation in technological dialogue and multi-national research projects. The President's agenda captures a balanced planning strategy for safe utilization of domestic fuels, and continued alternative energy research. And, the agenda is clearly supportive of, and actively engaged in international technologic exchanges. Altogether, these key points support a positive direction towards the reduction of fossil fuel reliance, and encourage all-aspects of clean-energy solutions.

Final Remarks

All U.S. citizens have a vested interest in clean and alternative energy solutions, and in our national energy direction. It is of substantial importance that we, as a society, and on global scale, work toward goals which will significantly reduce our reliance on fossil fuels. An important, essential step in decreasing the dependency on fossil fuels is continued scientific research, and exploitation of alternative energies. The protection of human health, our cultural identities and our environment (including wildlife, and habitats) is increasingly at risk. Furthering the use of alternative energies while maintaining the fundamentals of environmental protection, must be a societal goal. Importantly, the objectives to this goal must be promoted by effective leadership in government and industry, strengthened by the active participation of science, and combined with intense collaborative efforts. There is progress in many areas of energy research and application; however our society is at its infancy in developing effective, sustainable and environmentally safe alternative energies.

How can we do better and what more can society do to forge ahead smartly? I believe that one of society's primary fallbacks is the lack of complete awareness of the negative impacts from fossil fuel dependency. Although society in general realizes that energy from fossil fuels is a key issue today and for the future, the negative impacts are not given due appreciation; except in short-term cases when we are hit in the pocket-book by another round of high gas prices, or when we suffer a fuel-related tragedy such as that experienced by the *Deepwater Horizon* explosion. Then, when the cost of gas settles down, and no fuel related tragedy is being highlighted on television (because a leak has "finally" been stopped and the spill "somewhat" sponged up), society settles back into daily schedules, overlooking the negative impacts of fossil fuel dependency. Even the government appears to "get on" with other issues such as a current military action, election campaigns or investigating a federal scandal. As the events that highlight those negative impacts of fossil fuel dependency fade away momentarily, we start enjoying the comfort side of fossil fuels. Without much thought we fill up the gas tank while simultaneously checking for incoming texts, no longer concerned about the price since it has temporarily settled down to around 20 cents lower per gallon. Or, conveniently we enjoy the stadium lighting at a night game without wondering about the coal that is being burned to produce the electricity. Then, perhaps the negative aspects of fossil fuels are again dismissed when we are able to take in a breath of clean air and drink untainted water from the tap. Yes, sometimes society is "selectively" aware of the negative impact of fossil fuels, yet unless an event is affecting immediate comforts, or is being exploited on the news as a headline, many times we just move along entertaining all the "other issues" of the day. The public needs to be more aware of the negative impacts of fossil fuels to our lives and to the environment. Leaders of government, scholastic institutions and science need to help keep that awareness fresh and current in our minds, and offer direction and smart planning strategies.

To be fair, there are many individuals and organizations in society that are highly aware of the impact of fossil fuels. For example petroleum industry owners and employees, scientists, politicians, and wall-street executives, etc., have key interests in the impacts, status, and availability of fossil fuels. The industry owners and employees have vested interest in fossil fuels because that is how their living is made, whether it is from drilling, sales from infrastructure supplies, or working on the docks managing merchant tanker deliveries from foreign imports. Politicians have vested interests due to the outcry of the public to "lower" the prices at the pump and to make sure "we' have clean air and water. As well, politicians have a vested interest when their local district demands action because their community is being overridden with oil pumps and pipelines. In these cases, the demands will also assuredly encompass public concern of environmental destruction and contamination, as well as impacts to wildlife and sea life. As the nation's elected leader, the President has an overwhelming vested interest in changing society's dependency on fossil fuels, and must balance decisions with economic stability and negative environmental impacts. Scientists have vested interests because research and technology is the heart of all problem-solving prompting progress to future energy paths. Science will study the impacts to society and the environment, and will develop advanced technology to lesson impacts of energy sources, and to hone current processes. We all have a vested interest, and must have clear understanding of the impacts of our decisions.

Finally, it essential that science, industry, and government leaders develop relationships which will strengthen communications, knowledge and the understanding of the nation's energy needs and resources. It is imperative that effective alternative energies are developed that will decrease society's dependency on fossil fuels. When society fully understands the negative impacts of fossil fuels and applies the technology and processes of smart alternative energies, then human health, cultural history, and the precious environment will be the fortunate benefactors.

References

Air Products. (2013). News *Release: Air Products Celebrates Texas Carbon Capture Demonstration Project Achievement*. Retrieved from http://www.airproducts.com/company/news-center/2013/05/0510-air-products-celebrates-texas-carbon-capture-demonstration-project-achievement.aspx

Anderson, B. (2013). In Kodiak, moving Kulluk can wait until Tanner Crab Season Ends. *Alaskan Dispatch online*. Retrieved from http://www.alaskadispatch.com

Barcott, B. (2010). Forlorn in the Bayou. *National Geographic*, Vol. 218 Issue 4, p62-75. Retrieved from http://ehis.ebscohost.com.vlib.excelsior.edu

Biello, D. (2010). Lasting Menace. *Scientific American*, *303*(1), 16-18. Retrieved from http://web.ebscohost.com.vlib.excelsior.edu

Blake, A., Wade, A. (2013). USGS Releases New Oil and Gas Assessment for Bakken and Three Forks Formations – Finds Formations Have Greater Resource Potential than Previously Thought. *U.S. Department of the Interior News Release*. Retrieved from http://www.doi.gov/news/pressreleases/usgs-releases-new-oil-and-gas-assessment-for-bakken-and-three-forks-formations.cfm#

Boyd, D. (2012). Soaring Oil Production Spurs Infrastructure Growth Across Booming Bakken Play. *The American Oil & Gas Reporter*. Retrieved from http://www.aogr.com/index.php/magazine/cover-story/soaring-oil-production-spurs-infrastructure-growth-across-booming-bakken-pl

Brunner, J. (1999). Bellingham, Wash., to Clean Polluted Stream In Time for Salmon. *The Seattle Times (WA)*. Retrieved from http://ehis.ebscohost.com.vlib.excelsior.edu

Calhoun County Public Health Department. (2013). *Latest Enbridge News*. Retrieved from http://www.calhouncountymi.gov/government/health_department

Carbon Sequestration Leadership Forum (CSLF). (2013). A *Global Response to the Challenge of Climate Change*. Retrieved from http://www.cslforum.org/

Center for Disease Control (CDC). (2011). *Benzene, CAS ID #:71-43-2*. Toxic Substances Portal. Agency for Toxic Substances & Disease Registry. Retrieved from http://www.atsdr.cdc.gov/substances/toxsubstance.asp?toxid=14

Center for Disease Control and Prevention. (2010). *What Health Care Providers Should Know about Potential Health Hazards from the Deepwater Horizon Oil Spill*. Retrieved from http://www.bt.cdc.gov/gulfoilspill2010/pdf/Oil_spill_4p_July_29_v2.pdf

Charpentier, R., Cook, T., Gaswirth, S., Gautier, D., Higley, D., Klett, T., Lewan, M., Lillis, P., Marra, K., Schenk, C., Tennyson, M., and Whidden, K. (2013). *Assessment of undiscovered oil resources in the Bakken and Three Forks Formations, Williston Basin Province, Montana, North Dakota, and South Dakota, 2013*: U.S. Geological Survey Fact Sheet 2013–3013, 4 p. Retrieved from http://pubs.usgs.gov/fs/2013/3013/

Coleman, J., Demas, A., Pierce, B. (2011). *USGS Releases New Assessment of Gas Resources in the Marcellus Shale, Appalachian Basin*. USGS, Newsroom. Retrieved from http://www.usgs.gov/newsroom/article.asp?ID=2893#.UZGNbdDD-00

Courselle, D. (2010). We (Used To?) Make a Good Gumbo - The BP Deepwater Horizon Disaster and the Heightened Threats to the Unique Cultural Communities of the Louisiana Gulf Coast. *Tulane Environmental Law Journal, 24*19. Retrieved from http://ehis.ebscohost.com.vlib.excelsior.edu

Court, C., Jackson, R., White, N. (2012). The Role of Regional Science in Shale Energy Development. *Review of Regional Studies*, *42*(2), 176-105. Retrieved from http://ehis.ebscohost.com.vlib.excelsior.edu/eds/pdfviewer/pdfviewer?sid=79cb786c-5967-4b02-98f4-55c1e132c187%40sessionmgr104&vid=5&hid=101

Darrah, T., Down, A., Jackson, R., Karr, J., Osborn, S., Poreda, R., Vengosh, A., Warner, N., Zhao, K. (2013). *Increased Stray Gas Abundance in a Subset of Drinking Water Wells near Marcellus Shale Gas Extraction*. Proceedings of the National Academy of Sciences. Retrieved from http://www.pnas.org/content/early/2013/06/19/1221635110.full.pdf

Dobb, E. (2013). Bakken Shale Oil, The New Landscape. *National Geographic Magazine*. Retrieved from http://ngm.nationalgeographic.com/2013/03/bakken-shale-oil/dobb-text

Dow, K., Dr., Tuler, S., Dr., Webler, T., Dr. (2010). *Exclusive: The Human Dimensions of Oil Spills (2010)*. Retrieved from http://thinkprogress.org/climate

Energy Information Administration. (2015). *Annual Energy Outlook 2015*. Retrieved from http://www.eia.gov/forecasts/aeo/

Energy Information Administration. (2011). *Natural Gas, Maps: Exploration, Resources, Reserves and Production. Shale Gas and Oil Plays, Lower 48 States*. Retrieved from http://www.eia.gov/pub/oil_gas/natural_gas/analysis_publications/maps/maps.htm

Energy Information Administration. (2012). *U.S. Natural Gas Imports & Exports 2011*. Retrieved from http://www.eia.gov/naturalgas/importsexports/annual/

Energy Information Administration. (2013). *What are the Products and uses of Petroleum?* Retrieved from http://www.eia.gov/tools/faqs/faq.cfm?id=41&t=6

Environmental Protection Agency (EPA). (2012). *Summary of Past and Current EPA Response Activities Regarding the TVA Kingston Coal Ash Spill*. Retrieved from http://epa.gov/region4/kingston/summary.html

Environmental Protection Agency. (2013). *EPA's Study of Hydraulic Fracturing and its Potential Impact on Drinking Water Resources*. Retrieved from http://www2.epa.gov/hfstudy

Environmental Protection Agency. (2012). *EPA's Study of Hydraulic Fracturing and its Potential Impact on Drinking Water Resources: Progress Report*. Retrieved from http://www2.epa.gov/hfstudy/study-potential-impacts-hydraulic-fracturing-drinking-water-resources-progress-report-0

Environmental Protection Agency. (2013). *Natural Gas Extraction – Hydraulic Fracturing. Providing Regulatory Clarity and Protections against Known Risks*. Retrieved from http://www2.epa.gov/hydraulicfracturing

Exxon Valdez Oil Spill Trustee Council. (2010). *Exxon Valdez Oil Spill Restoration Plan. 2010 Update Injured Resources and Services, May 14, 2010.* Exxon Valdez Oil Spill Trustee Council, Anchorage, Alaska. Retrieved from http://www.evostc.state.ak.us

Fisk, M. (2013). BP Still Uncertain Over Spill Cost at Third Anniversary. *Bloomberg News*, Retrieved from http://www.bloomberg.com/news/2013-04-19/bp-still-uncertain-over-spill-cost-at-third-anniversary.html

Gill, D., Picou, J., Ritchie, L. (2012). The Exxon Valdez and BP Oil Spills: A Comparison of Initial Social and Psychological Impacts. *American Behavioral Scientist, 56*(1), 3-23. Retrieved from http://ehis.ebscohost.com.vlib.excelsior.edu

Hamilton, H., Slonecker, T. (2012). Measuring Landscape Disturbance of Gas Exploration in Bradford and Washington Counties. *USGS Newsroom*. Retrieved from http://www.usgs.gov/newsroom/article.asp?ID=3392#.UZGI1dDD-00

Hyder, M. (2011). Enbridge Spill Health Effects Surveyed. *Oil Spill Intelligence Report*, *34*(4), 2-3. Retrieved from http://ehis.ebscohost.com.vlib.excelsior.edu

Hydraulic Fracturing. (2013). Overview. *U.S. Geological Survey (USGS)*. Retrieved from http://energy.usgs.gov/OilGas/UnconventionalOilGas/HydraulicFracturing.aspx

Institute for Energy Research (IER). (2015). *Fossil Fuels*. Retrieved from http://www.instituteforenergyresearch.org/energy-overview/fossil-fuels/

Juhasz, A. (2012). Two Years Later: BP'S Toxic Legacy. Special Investigation. *Nation*, *294*(19), 11-15. Retrieved from http://ehis.ebscohost.com.vlib.excelsior.edu

Kerry, J. (2013). Remarks at Energy Cooperation Event. U.S. Department of State. Retrieved from http://www.state.gov/secretary/remarks/2013/04/207474.htm

Landers, J. (2009). *Tennessee Coal Ash Spill Prompts EPA Action, Separate Investigations*. Civil Engineering (08857024), 79(4), 24-27. Retrieved from http://ehis.ebscohost.com.vlib.excelsior.edu

LePoire, D. (2011). Exploring New Alternative Energy. *Futurist*, *45*(5), 34-38. Retrieved from http://ehis.ebscohost.com.vlib.excelsior.edu/eds

Lavelle, M. (2010). Special Report: The Great Shale Gas Rush. *National Geographic News*. Retrieved from http://news.nationalgeographic.com/news//2010/10/101022-energy-marcellus-shale-gas-environment/

McLeman, R. (2011). Settlement abandonment in the context of global environmental change. *Global Environmental Change, 21*(Supplement 1), Retrieved from http://www.sciencedirect.com.vlib.excelsior.edu/science

Mooney, C. (2011). The Truth about Fracking. *Scientific American*, 305(5), 80-85. Retrieved from http://ehis.ebscohost.com.vlib.excelsior.edu/eds/detail?vid=2&sid=d66e5865-c380-4c01-9866-281596d3d7ca

Muller, R. (2012). *Energy For Future Presidents: The Science Behind the Headlines.* New York, NY: W.W. Norton & Company, Inc.

National Renewable Energy Laboratory. (2013). *NREL Teams with Navy, Private Industry to Make Jet Fuel from Switchgrass.* Retrieved from http://www.nrel.gov/news/press/2013/2215.html

National Wildlife Federation. (2013). *Oil Spill Impacts on Fish and Aquatic Invertebrates.* National Wildlife Federation online. Retrieved from http://www.nwf.org/What-We-Do/Protect-Habitat/Gulf-Restoration/Oil-Spill/Effects-on-Wildlife/Fish.aspx

NOAA, Damage Assessment, Remediation Restoration Program (DARRP). (2013). *Case: Enbridge Pipeline Release.* Retrieved from http://www.darrp.noaa.gov/greatlakes/enbridge/index.html

NOAA, Office of Response and Restoration. (2013). *$44Million Natural Resource Damage Settlement to Restore San Francisco Bay After Cosco Busan Oil Spill (revised Jan 2013).* Retrieved from http://response.restoration.noaa.gov/about/media/44-million-natural-resource-damage-settlement-restore-san-francisco-bay-after-cosco-busa

NOAA, Office of Response and Restoration. (2013). *Case: Whatcom Creek, WA (revised Jan 2013).* Retrieved from http://www.darrp.noaa.gov/northwest/olympic/index.html

NOAA. (2012). *Gulf Spill Restoration. Affected Gulf Resources*. April 2012 Natural Resource Damage Assessment Status Update for the Deepwater Horizon Oil Spill. Retrieved from http://www.gulfspillrestoration.noaa.gov/oil-spill/affected-gulf-resources/

NOAA, Office of Response and Restoration. (2013). *NOAA Responds to Shell Drilling Rig Kulluk Grounding in Gulf of Alaska.* Retrieved from http://response.restoration.noaa.gov/about/media/noaa-responds-shell-drilling-rig-kulluk-grounding-gulf-alaska.html

NOAA, Damage Assessment, Remediation, & Restoration Program. (2002). *Restoration Plan and Environmental Assessment for the June 10, 1999 Olympic Pipeline Gasoline Spill into Whatcom Creek, Bellingham, Washington*. Retrieved from file 1 at http://www.darrp.noaa.gov/northwest/olympic/admin.html

Picou, J. (2009). When the Solution Becomes the Problem: The Impacts of Adversarial Litigation on Survivors of the Exxon Valdez Oil Spill. *University Of St. Thomas Law Journal: Fides Et Lustitia*, 768. Retrieved from http://ehis.ebscohost.com.vlib.excelsior.edu

Plume, K. (2013). *Mississippi River Barge Backup shrinks, Oil Cleanup Continues.* Retrieved from http://www.reuters.com/article

Substance Abuse and Mental Health Services Administration (SAMHSA). (2013). *Behavioral Health in the Gulf Coast Region Following the Deepwater Horizon Oil Spill*. Center for Disease Control and Prevention. Retrieved from http://www.samhsa.gov/data/nsduh/nsduh-gsps-gulf-coast.pdf

Tennessee Department of Health. (2010). *Public Health Assessment Final Release*. Tennessee Valley Authority (TVA) Kingston Fossil Fuel Plant. Retrieved from http://www.cdc.gov

U.S. Coast Guard Eighth District External Affairs. (2013). Coast *Guard Responds to Report of Crude Oil in Mississippi River*. Retrieved from http://www.uscgnews.com/go/doc/4007/1689811

U.S. Department of Energy. (2011). *U.S.-China Clean Energy Cooperation. Progress Report*. Retrieved from http://energy.gov/sites/prod/files/piprod/documents/USChinaCleanEnergy.PDF

U.S. Department of State. (2013). *U.S.-India Joint Fact Sheet: Sustainable Growth, Energy and Climate Change*. Retrieved from http://www.state.gov/r/pa/prs/ps/2013/06/211017.htm

U.S. Energy Department, Office of Fossil Energy. (2013). *United States-United Kingdom Collaboration on Fossil Fuel Energy R&D* (2013). Retrieved from http://energy.gov/fe/united-states-united-kingdom-collaboration-fossil-energy-rd

U.S. Energy Information Administration (EIA). (2012). *How Dependent are we on Foreign Oil?* Retrieved from http://www.eia.gov/petroleum/

U.S. Energy Information Administration (EIA). (2015). *How many and what kind of power plants are there in the United States?* Retrieved from http://www.eia.gov/coal/

U.S. Energy Information Administration (EIA). (2015). *What are the products and uses of petroleum?* Retrieved from http://www.eia.gov/tools/faqs/faq.cfm?id=41&t=6

U.S. ITER. (2012). *Background*. Retrieved from https://www.usiter.org/about/

Whitehouse. (2011). *Blueprint for a Secure Energy Future*. Retrieved from http://www.whitehouse.gov/energy

Whitehouse. (2013). *Energy, Climate Change and Our Environment*. Energy section. Retrieved from http://www.whitehouse.gov/energy

www.ingramcontent.com/pod-product-compliance
Lightning Source LLC
Chambersburg PA
CBHW040906180526
45159CB00010BA/2944